Daniel Augustus Beaufort

Memoir of a Map of Ireland;

Illustrating the Topography of that Kingdom, and Containing a Short....

Daniel Augustus Beaufort

Memoir of a Map of Ireland;
Illustrating the Topography of that Kingdom, and Containing a Short....

ISBN/EAN: 9783337184452

Printed in Europe, USA, Canada, Australia, Japan

Cover: Foto ©berggeist007 / pixelio.de

More available books at **www.hansebooks.com**

MEMOIR

OF A

MAP OF IRELAND;

ILLUSTRATING

THE TOPOGRAPHY OF THAT KINGDOM,

AND CONTAINING

A SHORT ACCOUNT OF ITS PRESENT STATE,

Civil and Ecclesiastical;

WITH

A COMPLETE INDEX TO THE MAP.

By *DANIEL AUGUSTUS BEAUFORT, L.L.D.*

RECTOR OF NAVAN, IN THE COUNTY OF MEATH, AND VICAR OF
COLLON, IN THE COUNTY OF LOUTH.—M. R. I. A.

- - - - - - Situ ac falubritate cœli atque temperie, acceſſu cunctarum gentium
facili, littoribus portuoſis, aquarum copia, montium articulis, ferorum animalium
innocentia, ſoli fertilitate, pabuli ubertate: quicquid eſt quo carere vita non debeat,
nuſquam eſt præſtantius; fruges, vellera, lina, juvenci.

PLIN. HIST. NAT. Lib. 37.

LONDON:

SOLD BY W. FADEN, GEOGRAPHER TO THE KING, CHARING-CROSS; J. DEBRETT,
PICCADILLY, AND JAMES EDWARDS, PALL-MALL.

1792.

THE KING.

SIR,

THE fplendid acquifitions that have been made to remote Geography, under Your MAJESTY's immediate aufpices and protection, and the liberal and enlightened patronage You have extended to every attempt towards the improvement of this ufeful fcience, embolden me to lay at your MAJESTY's feet, this humble endeavour to elucidate and amend the Topography of fo refpectable and important a part of the Britifh Empire, as the Kingdom of IRELAND.

Your

DEDICATION.

Your MAJESTY's paternal attention to the interefts of the National Church ftill farther encourages me to hope, that You will deign to receive favorably this FIRST attempt to trace the ecclefiaftical divifions, and to delineate the diocefan diftricts, of an entire Kingdom.

May your MAJESTY long continue to promote the caufe of religion and virtue, by the due exercife of your authority, and the influence of your Royal Example.

I am,

SIR,

With the higheft refpect,

Your MAJESTY's

Moft dutiful fubject and fervant,

Daniel Auguftus Beaufort.

PATRONISERS and SUBSCRIBERS

M A P.

THE KING.

His Royal Highnefs the PRINCE OF WALES.

His Royal Highnefs the DUKE OF GLOUCESTER.

His Excellency the EARL OF WESTMORELAND,
LORD LIEUTENANT OF IRELAND.

A

His Grace Richard *Lord Rokeby*, Archbifhop of Armagh, Lord Primate of Ireland

Rt. Hon. Earl of Aylefbury, K. T. *Lord Chamberlain to the Queen*

Rt. Hon. Lord Vifcount Allen

Rt. Hon. Lord Auckland, *Ambaffador Extraordinary to the States of Holland, F. R. S.*

Henry Alexander, Efq. M. P.

Rev. Richard Allot, D. D. *Precentor of Armagh, and Treafurer of Chrift Church, M. R. I. A.*

William Anderfon, Efq.

Rev. H. Annefley

Henry Arabin, Efq.

Rev. Nicholas Afhe

Rev. St. George Afhe

Rev. Gilbert Auftin, *M. R. I. A.*

B

His Grace the Duke of Buccleugh, K. T.

Rt. Hon. Marquis of Buckingham, K. G.

Rt. Hon. Earl of Briftol, *Lord Bifhop of Derry, F. R. S.*

Rt. Hon. Lord Vifcount Belmore

Rt. Hon. John Beresford, *Firft Commiffioner of the Revenue*

Rt. Hon. Lieut. Gen. John Burgoyne

Hon. Rev. Charles Broderick, *Treaf. of Cloyne*

Hon. Mr. Juftice Boyd

Sir Jofeph Banks, Bart. *Pref. R. S. and M. R. I. A.*

William Bagot, Efq.

Rev. John Bailey

Rev. John Ball

Frederick Barnard, Efq. *F. R. S.*

Rev. John Barry, D. D. *Dean of Elphin*

Gaynor Barry, Efq.

William Bellingham, Efq. *Commiffioner of the Navy*

Rev. Conway Benning, LL.D.

Rev. Hill Benfon

Richard Benyon, Efq.

Robert Bermingham, Efq.

Henry Bingham, Efq.

Rev.

Rev. Stewart Blacker, *Dean of Leighlin and Archdeacon of Dromore*
Rev. Henry Blacker
Mrs. Blenerhaffet, Bath
Mrs. Letitia Blenerhaffet
Arthur Blenerhaffet, Efq. *Bath*
Arthur Blenerhaffet, Efq. *Arbela*
Arthur Blenerhaffet, Efq. *Tralce*
William Blenerhaffet, Efq.
Samuel Blount, Efq.
George Bogg, Efq.
Cornelius Bolton, Efq. *M. R. I. A.*
Rev. Richard Bourne, *Chancellor of St. Patrick's, Dublin*
James Bradifh, Efq.
Mifs Catherine Beaver Browne
Rev. Jemmet Brown
Dodwell Browne, Efq:
Rev. Thomas Brownrigg
Rev. Theobald Brownrigg
John Brownrigg, Efq.
Mr. John Brownrigg
Rev. John Broughan
Rev. John Buck, *M. R. I. A.*
Edward Bulkely, Efq.
Richard Burden, Efq.
Thomas Burgh, Efq. *M. P.*
Beresford Burfton, Efq.
Charles William Bury, Efq. *M.P.* and *M. R. I. A.*
Pierce Archer Butler, Efq.

C

His Grace Dr. *John Moore*, Lord Archbifhop of Canterbury
His Grace Dr. *Charles Agar*, Lord Archbifhop of Cafhel, and *M. R. I. A.*
Rt. Hon. Earl of Cardigan
Rt. Hon. Earl of Carlifle, *K. T.*
Rt. Hon. Earl of Chatham, *K. G. Firft Lord of the Admiralty*

Rt. Hon. Earl of Courtown, *K. St. P. Treafurer of the King's Houfhold*
Rt. Hon. Earl of Charlemount, *K. St. P. Pref. R. I. A. and F. R. S.*
Rt. Rev. Dr. *John Hotham*, Lord Bifhop of Clogher
Rt. Rev. Dr. *Richard Woodward*, Lord Bifhop of Cloyne
Rt. Rev. Dr. *William Bennet*, Lord Bifhop of Cork
Rt. Hon. Lord Cloncurry
Rt. Hon. James Cuffe, *M. P.*
Hon. Rev. Hamilton Cuffe
Hon. Rev. Maurice Crofbie, *D.D. Dean of Limerick*
Jofeph Calcut, Efq.
Andrew Caldwell, Efq. *M. R. I. A.*
Turner Camac, Efq.
Rev. Drelincourt Campbell
Rev. Thomas Campbell, *LL.D. Chancellor of Clogher*
Oliver Carleton, Efq.
Rev. Peter Carleton, *Dean of Killaloe*
Rev. John Carpendale, *D.D.*
Rev. Jofeph Caffan
Rev. C. Byrne Caulfield, *Archdeacon of Clogher*
James Caulfield, Efq.
Thomas Caulfield, Efq.
Mr. John Chambers
Rev. Arthur Champagne, *Dean of Clonmacnoife*
Mr. Mathew Charley
Rev. Jofeph Clarke
Maxwell Clofe, Efq:
Rev. John Connor, *D. D.*
John Cooke, Efq. *M. R. I. A.*
Auftin Cooper, Efq.
John Cooper, Efq.
Samuel Cooper, Efq.
Robert Cornwall, Efq.
Rev. James Cottingham, *D.D. Vicar General of Kilmore*

Thomas

Thomas Cowan, Efq:

Rev. Marmaduke Cramer, *D. D. Chancellor of Chrift Church*

Marmaduke Cramer, Efq.

Rev. Cecil Crampton

Morgan Crofton, Efq.

Hugh Crofton, Efq.

Rev. Henry Crofton

Rev. John Cromie

Edward Crofbie, Efq.

Mr. John Crofthwaite, *M. R. I. A.*

Rev. Dawfon Crowe, *D. D.*

Rev. William Crowe

D

His Grace Dr. *Robert Fowler*, Lord Archbifhop of Dublin

His Grace the Duke of Devonfhire, *K. G.*

Rt. Rev. Dr. *Thomas Percy*, Lord Bifhop of Dromore, *M. R. I. A.*

Rt. Rev. Dr. *William Dickfon*, Lord Bifhop of Down

Rt. Hon. Marquis of Downfhire, *F. R. S.*

Sir Charles Defvœux, Bart. M. P.

Marriot Dalway, Efq.

John Daly, Efq.

John Darcy, Efq.

Arthur Dawfon, Efq.

Rev. Robert Dealtry, *LL.D. Precentor of St. Patrick's, Dublin*

John De Courcey, Efq.

Rev. Francis Defpard

Rev. William Digby, *Dean of Clonfert*

Rev. Simon Digby

Gordon Dillingham, Efq. *Letton, Norfolk*

Rev. Richard Dobbs, *Dean of Connor, M. R. I. A.*

Conway Richard Dobbs, Efq.

Edward Price Dobbs, Efq.

Francis Dobbs, Efq.

Rev. Oliver Dodd

James Donaldfon, Efq.

Rev. Dive Downes, *D. D.*

Rev. Clotworthy Downing

William Doyle, *LL.D. M. R. I. A.*

Major John Doyle, *Secretary to His Royal Highnefs the Prince of Wales*

Rev. James Drought, *D. D.*

Patrick Duigenan, *LL.D. Vicar General of Dublin, &c.*

Francis Durour, Efq.

Rev. Philip Daval, *D. D. Canon of Windfor, F. R. S.*

E

Right Rev. *Dr. Charles Dodgfon*, Lord Bifhop of Elphin, *F. R. S.*

Richard Eaton, Efq.

Mrs. Echlin

Rich. L. Edgeworth, Efq. *F.R.S. M.R.I.A.*

Robert Ellis, Efq.

Rev. Thomas Ellifon

Rev. John Ellifon, *D. D.*

Rev. Thomas Elrington, *Fellow of Trinity College, Dublin, M. R. I. A.*

F

Rt. Hon. Earl Fitzwilliam

Rt. Hon. Earl of Fauconberg

Rt. Hon. Earl of Farnham

Rt. Rev. *Dr. Eufeby Cleaver*, Lord Bifhop of Ferns

Rt. Hon. John Fofter, *Speaker of the Houfe of Commons of Ireland, M. R. I. A.*

Rt. Hon. Colonel Richard Fitzpatrick

Rt. Hon. Charles James Fox

Sir John Freke, Bart. M. P.

Maximilian Faviere, Efq.

Anthony Fergufon, Efq.

Mrs. Catherine Finn

Rev. Gerald Fitzgerald, *D. D. Senior Fellow of Trinity College, Dublin*

Richard Frankland, Efq. *M. R. I. A.*

Robert French, Efq.

Rt.

G

Rt. Hon. Earl of Glandore, *M. R. I. A.*
Nicholas Gay, Efq.
Rev. Daniel Gealy
Rev. John Gibbons
Rev. Wood Gibfon, *D. D.*
John Godley, Efq.
Thomas Goold, Efq.
George Goflling, Efq.
Richard Gough, Efq. *F. R. S.*
Standifh Grady, Efq.
Rev. Thomas Grady
Hector Graham, Efq.
Rev. Thos. Graves, *D. D. Dean of Ardfert*
William Gray, *M. D.*
John Greer, Efq.
Richard Griffith, Efq. *M. R. I. A.*

H

Rt. Hon. Earl of Hillfborough, *F. R. S.*
Rt. Hon. John Hely Hutchinfon, M. P. *Secretary of State, and Provoft of Trinity College, Dublin, M. R. I. A.*
Sir Francis Hutchinfon, Bart. M. P.
Rev. William Hales, *D. D. M. R. I. A.*
Rev. George Hall, *D. D. Sen. Fell. Trin. Coll. Dub. M. R. I. A.*
Rev. Hugh Hamilton, *D. D. Dean of Armagh, F. R. S. and M. R. I. A.*
Rev. William Hamilton, *M. R. I. A.*
Rev. James Archibald Hamilton, *D. D. Archdeacon of Rofs, M. R. I. A.*
Rev. William Slicer Hamilton
Thomas K. Hannington, Efq.
Rev. Singleton Harpur
George Harris, *LL. D.*
Sir Henry Harftonge, Bart:
John Hatch, Efq.
Jofeph Henry, Efq. *M. R. I. A:*
Hugh Henry, Efq,
Henry Herbert, Efq,

James Heffeltine, Efq.
Lancelot Hill, Efq.
Rev. John Hill
Rev. Thomas Hore
Rev. John Humfrey, *Rector of Dunham, Norfolk*

I

Alexander Jaffray, Efq. *M. R. I. A:*
Rev. Samuel Johnfon
Rev. John Jones
Rev. William H. Irvine

K

Rt. Hon. Earl of Kingfton
Rt. Hon. Lord Vifcount Kingfland
Rt. Rev. *Dr. G. L. Jones*, Lord Bifhop of Kildare and Dean of Chrift Church, *M. R. I. A.*
Rt. Rev. *Dr. Thomas Barnard*, Lord Bifhop of Killaloe, *F. R. S. and M. R. I. A.*
Rt. Rev. *Dr. John Law*, Lord Bifhop of Killalla, *F. R. S. and M. R. I. A.*
Rt. Rev. *Dr. William Fofter*, Lord Bifhop of Kilmore
Mr. Valentine Kealy
Rev. Michael Kearney, *D. D. M. R. I. A.*
Rev. John Kearney, *D. D. Sen. Fell. of Trin. Coll. Dub. M. R. I. A.*
Rev. Cadogan Keating, *Dean of Clogher*
John Kelly, Efq.
Rev. John Kenny, *Vicar General of Cork*
Rev. Francis Kenny
Rev. John King, *D. D.*
Rev. George Knox

L

Library of Trinity College Dublin
Rt. Hon. Marquis of Lanfdown, *K. G.*
Rt. Rev. *Lord Glenworth*, Lord Bifhop of Limerick

John

John Ladeveze, Efq.
Rev. George Lambart
Edward Law, Efq.
James Law, Efq.
Rev. Edward Ledwich, *M. R. I. A.*
Rev. Philip Lefanu, *D. D. M. R. I. A.*
Leut. Col. Anthony Lefroy, *M. R. I. A.*
Thomas Legge, Efq.
Captain Legge
John Leigh, Efq.
Charles Powel Leflie, Efq. M. P.
Rev. Henry Leflie, *D. D.*
Rev. John Letablere
Robert Lindfay, Efq.
Rev. James Little, *D. D.*
Rev. William Lodge, *LL. D.*
John Longfield, *M. D.*
Francis Longworth, Efq.
Rev. Verney Lovett, *M. R. I. A.*
Nathaniel Low, Efq.
John Lyfter, Efq.

M

Rt. Hon. Earl of Moira, *F. R. S. and M. R. I. A.*
Rt. Hon. Lord Vifcount Mountmorris, *M. R. I. A.*
Rt. Hon. Lord Vifcount Maxwell
Rt. Rev. *and Hon. Dr. Henry Maxwell*, Lord Bifhop of Meath
Rt. Hon. Lord Macartney, *K. B.*
Lady G. Meredyth.
Sir Richard Gorges Meredyth, *Bart.*
John Macartney, Efq.
John Macaulay, Efq. *M. R. I. A.*
Rev. John M'Caufland
Rev. William M'Cleverty
John M'Clintock, Efq. *M. P.*
Hugh M'Dermott, *M. D.*
Daniel M'Gufty, Efq.
Robert, Macky, Efq.
Daniel Macnamara, Efq.
Mr. John Magee
Rev. Thomas Mahon

Mrs. Mahon
Edmond Malone, Efq. *M. R. I. A.*
Rev. Robert Marfh
Rev. Jeremiah Marfh
Francis Marfh, Efq.
Mifs Maffingberd
Peter Maturin, Efq.
John Maxwell, Efq.
Robert Maxwell, Efq.
Rev. James Maxwell
Robert Mayne, Efq.
Major John Mercier
Mrs. Meredyth
Rev. Alexander Montgomery
John Montgomery, Efq.
Nathaniel Montgomery Moore, Efq.
Rev. Richard Murray, *D. D. Vice. Prov. of Trin. Coll. Dublin, M. R. I. A.*

N

Mr. Deputy John Nicholls
Mrs. Nefbitt
Mrs. Norris

O

Rt. Rev. *and Hen. William Berefford*, Lord Bifhop of Offory
Dennis O'Connor, Efq.
John O'Donnell, Efq.
James Ormfby, Efq.
John Ormfby, Efq.
Mr. Andrew Ormfton
Rev. John Orr
Ralph Oufley, Efq.

P

His Grace the Duke of Portland, *F. R. S.*
Rt. Hon. William Pitt, *Chancellor of the Exchequer, &c. &c. &c.*
Rt. Hon. Sir John Parnell, *Bart. Chancellor of the Exchequer of Ireland*
Hon. Rev. John Pomeroy
Rev. Thomas Pack, *Dean of Offory*

B   Rev.

PATRONISERS AND SUBSCRIBERS

Rev. Anthony Pack
Rev. Thomas Pack
Roger Palmer, Esq.
Robert P. Parker, Esq.
Rev. John Parker, *D. D. Vicar General of Killaloe*
Rev. Thomas Paul, *LL.D.*
Rev. Mr. Peel, *Norwich*
Robert Percival, *M. D.—M. R. I. A.*
Edward Piers, Esq.
Dowdall Pigot, Esq.
William Moreton Pitt, Esq *M.P. and F.R.S.*
John Pollock, Esq.
Simon Preston, Esq.

Q

Rev. John Quin

R

Rt. Rev. *Dr. James Hawkins*, Lord Bishop of Raphoe
Andrew Ram, Esq.
Stephen Ram, Esq.
Rev. Edward Sterling Roberts
Jonathan Boyle Roberts, Esq.
Rev. Thomas Robinson, *D. D.*
Rev. George Rogers, *A. M. Chancellor of Dromore*
William Rogers, Esq.
Rev. Samuel Ryall
Rev. Dudley Ryder
Rev. Dudley Ryves

S

Rt. Hon. the Marquis of Salisbury, *F. R. S.*
Rt. Hon. Earl of Shannon, *K. S:. P.*
Rt. Hon. Lord Viscount Sidney
Hon. Rev. Thomas Stopford, *Dean of Ferns*
Sir Annesley Stewart, *Bart. M. P.*
Sir George Leonard Staunton, *Bart. F.R.S.*
Sir John Scott, *Solicitor General, M. P.*
Major General John Straton
Rev. Howard St. George, *D. D.*

Rev. B. St. George
Rev. Henry St. George
Rev. William Sandford
Patrick Savage, Esq.
Rev. John Scott
John Scott, Esq.
Rev. John Seymour
Rev. Martin Sherlock, *Archdeacon of Killalla*
Rev. Thomas Simcox
Mr. William Sleater
Michael Smith, Esq
Rev. Thomas Smyth, *D. D.*
Ralph Smyth, Esq.
Rev. Clotworthy Soden, *Archdeacon of Derry*
Robert Sproule, Esq.
Herbert Stepney, Esq.
James Stewart, Esq. *M. P.*
William Stewart, Esq.
John Stewart, Esq.
Alexander Thomas Stewart, Esq. *M. R. I. A.*
Rev. Hugh Stewart
Rev. I. Stewart
Rev. Joseph Stock, *D. D.*
Rev. Charles Stone, *D. D. Archdeacon of Meath*
Rev. James Stopford
Rev. Joseph Stopford
Rev. Joseph Story
Thomas Stoughton, Esq.
Rev. Mr. Sturrock
Maurice Swabey, Esq.
Rev. Samuel Synge, *Archdeacon of Killaloe*
Edward Synge, Esq.
Rev. Edward Synge
Rev. Edward Synge, jun.
Robert Synge, Esq.

T

His Grace *the Earl of Mayo*, Lord Archbishop of Tuam, *M. R. I. A.*

Daniel

Daniel Tighe, Efq.
Rev. Thomas Tighe
Michael Tifdale, Efq. *M. R. I. A.*
John Toler, Efq. *Solicitor General, M. P.*
Moft Rev. Dr. Troy
Rev. Jofeph Turner, *D. D. Dean of Norwich*
Rev. Peter Turpin

V

Rt. Hon. Lord Vifcount Valentia, *M.R.I.A.*
Francis Vefey, Efq.
John Vefey, Efq.
Rev. Thomas Vefey
Rev. John Vignoles
Richard Vincent, Efq.

W

Rt. Rev. *Dr. Wllliam Newcome*, Lord Bi-
fhop of Waterford, *M. R. I. A*
Sir George Alanfon Winn, *Bart. M. P.*
Jofeph Cooper Walker, Efq. *M. R. I. A.*

John Wallis, Efq.
Rev. Raphael Walfh, *Dean of Dromore*
Rev. Henry Lomax Walfh, *LL.D.*
Peter Walfh, Efq.
John Warren, Efq.
Nicholas Weftby, Efq. *M. P.*
William Wittingham, Efq.
Rev. James Whitelaw
Ezekiel David Williams, Efq.
Charles H. Wilfon, Efq.
Mr. William Wilfon
George Frederick Winflanley, Efq.
Rev. William Wolfeley
Rev. Charles Woodward, *D. D.*
Mrs. Judith Woodward
Rev. Henry Wynne
Rev. Richard Wynne

Y

Rev. Mathew Young, *D. D. S. F. Trin.*
Coll. Dublin, M. R. I. A.

CONTENTS.

PREFACE.

THE candour and liberality with which my propo-
pofals for this work have been received, call upon me
to apologize for the length of time that has elapfed
fince they were firft publifhed. The moft effectual
and fatisfactory method of doing this, will be, to relate
the motives which originally induced me to ftep afide
from my profeffional ftudies, into the province of the
geographer, and have finally determined me on offer-
ing to the public an *entire new Map* of the kingdom
of IRELAND.

The firft idea of the work was fuggefted by the
difficulties which I had often experienced, in endea-
vouring to trace out the ecclefiaftical divifion of the
kingdom, and to afcertain the limits of each diocefe.
Nothing can be more intricate than thofe divifions;
their boundaries and extent having little or no depend-
ance on thofe of counties and baronies. The clergy of
the county of Galway are under the patronage of five

different

different Sees; and thofe of the Queen's County, which in fize is not a fourth of Galway, depend upon as many; while the bifhops of Meath and Killaloe extend their jurifdiction into fix counties.

The ecclefiaftical and civil divifions of the kingdom being fo unaccountably intermixed; it occurred to me, that a Map of Ireland, in which the extent of every diocefe and the fite of their feveral parifhes fhould be af-certained, might be acceptable to the public: and I was tempted to undertake it. Having therefore communi-cated my intention to moft of the Bifhops, it was imme-diately honoured with their Lordfhips approbation; and, encouraged by their patronage, I proceeded in my defign.

But that defign went no farther, at firft, than to in-fert thefe particulars in a faithful copy of one of the beft and moft modern maps of this country: and it was not till after I had employed much time and pains on it, that I found the fcale which I had adopted, of ten miles to an inch, too contracted for my purpofe; and the maps which I intended to follow, fo full of errors and defects, as to require almoft perpetual cor-rection.

What was then to be done?—I could neither refolve to relinquifh my defign, nor fubmit to the publifhing a map crowded and full of faults.—I determined, there-fore,

fore, at once, to alter and enlarge my plan; thinking
that my leifure hours could not be better employed
than in correcting the geography of this kingdom.
For this end I was obliged to fet about my work, as
if no general map of Ireland had been extant: and
without paying the fmalleft attention to thofe of Moll,
Jeffereys, Kitchin, Rocque, Bowles, &c. I have *con-
ftructed a new one*, upon two fheets, by a fcale of fix
miles to an inch, from the beft authorities and moft
authentic information that I have been able to procure.

A perfectly correct map cannot be expected, till every
county has been accurately furveyed; and it is to be lamented, that the aftronomer and the engineer have been
fo much lefs employed in fettling the geography of the
Britifh Iflands, than in afcertaining that of our diftant
poffeffions. The coafts and harbours of India and
America are better known, and more correctly laid
down, than thofe of Ireland or even of Great Britain.

In the year 1655, SIR WILLIAM PETTY furveyed all
the forfeited lands in the kingdom; under which defcription the greater part of moft counties, and the whole of
others were included. Since that period, twelve counties only have been furveyed; and of thofe no more
than nine maps are publifhed. Thofe of Meath,
Donegal, and Tyrone, and the *new* map of Armagh,

I   continue

continue in manufcript, and are depofited in the refpective court-houfes, for the ufe of the gentlemen of
each county.

But as IRELAND is rapidly increafing in population and opulence, and liberally encourages every art
and fcience ; there can be little doubt, that the time is
not very diftant, when the topography of every part of
it will be completely afcertained.

Meanwhile I offer to the public this Map; in which
I have endeavoured to give a faithful reprefentation of
the face of the country; by delineating, with all the
precifion in my power, the courfe of rivers, the fituation
and comparative heights of mountains, and the relative
fize and confequence of towns and villages : and I flat-
ter myfelf that it will afford fome interefting information, not only to Irifhmen and Britons, but even to fo-
reigners.

The mountainous appearance of the weftern coafts
will account for the little ufe that is made of fome of
the fineft harbours in the world; the multitude of parifhes
in the eaftern and fouthern counties, prove thofe to have
been the moft wealthy and populous parts of the kingdom, at a very early period. Many political and hifto-
rical deductions, which may be drawn from the local
circumftances of a country, will ferve either to confirm

or

or refute its ancient and obfcure annals. Few countries in the world have experienced greater viciffitudes than Ireland; having been, if we may give credit to her hiftorians, populous and civilized, long before the days of St. Patrick; and having funk, afterwards, into a ftate of almoft favage ignorance and barbarifm. Inftances of fuch degeneracy are not frequent, and much has been written of late, with great ingenuity, to difprove the exiftence of that flourifhing ftate, which fhe is faid to have enjoyed in remote ages. Without entering here into a difquifition of fo nice a queftion; or attempting to decide between the zealous champions for her ancient grandeur, and the fceptical antiquaries, who endeavour, by learned arguments and reafoning, to confute long-credited traditions; I fhall only obferve, that if, in the warmth of national enthufiafm, one party feems to raife the glory of ancient Ireland too high; the other, perhaps, through an honeft indignation againft the legendary tales and fiction, that fo often fully the page of early hiftory, too much deprefs its former condition.—The truth may probably lie between them.—And an intimate acquaintance with the face of this country, joined to an attentive obfervation of the changes it has undergone, and of the various monuments of antiquity which ftill remain in

every

every part of it, is effentially neceffary to the philofo-
phic enquirer, and his fureft guide in the inveftigation
of its true hiftory, during the times which preceded
the reign of Henry II. and the three fucceeding ages.

But to return to the immediate fubject of this
memoir; I fhall now enumerate the fources from
which I have drawn my information, and the authorities upon which I have ventured to deviate from former geographers. The ufe I have made of thefe authorities, and the manner in which I have compared and
combined them, fhall be particularized in the following fheets.

Sir William Petty's Maps of Ireland, and of each of
its counties, firft made public in 1685, have been the
ground-work of mine, as well as of every other map
of this kingdom, that has been publifhed in the courfe
of this century.

The ECCLESIASTICAL part of my map is much in-
debted to them, but ftill more to his orignal furveys of
the feveral baronies and parifhes, which are preferved
as records, in the Surveyor-general's office, in the Caftle
of Dublin. To thefe I had free and frequent accefs
through the liberal and obliging permiffion of Mr. Hand-
cock, deputy furveyor-general.

I have

I have received confiderable affiftance from the map of the county of Down, publifhed without a name, in 1767; and from that of Lough Neagh and its environs, by Mr. J. Lendrick in 1785. His map of Antrim, pub-lifhed in 1780, and Mr. Neville's of Wicklow in 1760, with lieutenant Alexander Taylor's excellent and accu-rate maps of Kildare and Louth, the former in 1780, and the latter in 1787, have enabled me to give a very true reprefentation of thofe counties.

Rocque's Map of the county of Dublin has affo:ded much information. His Map of Armagh in 1760, and Oliver Sloane's of the Queen's County, engraved about thirty years ago, though very inferior performances, were however of fome ufe; and Mr. Pelham's recent furvey of Clare, fully eftablifhes the topography of that county.

To thefe muft be added an old furvey of Cork har-bour, publifhed by Mount and Page; one, of the har-bour of Waterford by William Doyle in 1735, and another by M. M'Kenzie in 1767 ; with Mr. Cowen's maps of the river Shannon, and Mr. N. Roche's of the Suir.

I have alfo confulted Dr. Smith's account of the ancient and prefent ftate of the counties of Cork, Wa-terford, and Kerry; and have derived great advantage,

in

in delineating the coafts, from Mr. Mackenzie's, and
Captain Huddart's charts.—Meffrs. Taylor and Skin-
ner's very exact and ufeful furvey of the roads of Ireland, has been of great fervice to me in many refpects,
but principally in afcertaining the diftances of towns
and reprefenting the face of the country.

Befide thefe printed authorities I have been favoured
with the ufe of fome excellent drawings.—To the
Grand Juries of Donegal and Tyrone, I am obliged,
for the liberty of reducing to the fcale of my map,
their large and elegant maps of thofe counties, which
were actually furveyed a few years ago, and I have
reafon to believe, with great fkill and accuracy, by
Meff. Macrea of Lifford.—I have been alfo indulged,
at Armagh, with a tranfcript of the new map of that
county, by the fame artifts.   Sloane's map of Meath,
which is on a large fcale, has fupplied me with the
relative pofitions of places in that county.

I have had the advantage of tracing the rivers Shannon, Boyne, and Brofna, from Mr. Bernard Scalé's
furveys of thofe rivers, by permiffion of Mr. Lof-
tus, and the commiffioners of impreft accounts,
in whofe office thofe plans are depofited.   Colonel
Tarrant obliged me with his furvey of the river

3   Barrow,

Barrow, and Mr. Cowen with his original drawings of the Shannon.

A furvey of the whole tract of country, through which a canal was fome years ago propofed to be made, from Dublin to the river Inny, was kindly communicated to me by William Smyth of Barbavilla Efq; and has affifted me in correcting the fituation of places and the courfe of rivers along that line ; through parts of the counties of Meath and Weftmeath.

The courfe of the grand and royal canals, as far as they are executed, I have received from the accurate pencil of their furveyor, Mr. John Brownrig. To the Rev. Mr. Whitelaw, and the Rev. Mr. Harvey, my grateful acknowledgements are due, for the unfolicited communications of their elegant maps of the baronies of Tirawly, in the county of Mayo, and of Inifhowen, in Donegal. Nor muft I omit to thank Lieutenant Taylor, who is actually engaged in furveying the county of Longford, for a fketch of the diftances and bearings of the feveral towns in that county; which he very obligingly communicated at the requeft of my learned and ingenious friend Richard Lovel Edgworth, Efq.

But nothing can more effectually contribute to rectify the geography, and to afcertain the figure and

c   extent

extent of a country, than the determining the latitude
of a number of places in it, by accurate aſtronomical
obſervations: and of theſe I have availed myſelf, whenever I could depend upon their preciſion.—Such are
thoſe which I received from a much lamented friend,
and excellent aſtronomer, the late Dr. Uſher, F. R. S.
and Profeſſor of aſtronomy in the Univerſity of Dublin;
and thoſe with which I was ſupplied, by the friendſhip
of the Rev. Dr. J. A. Hamilton of Armagh, and the
politeneſs of Dr. Longfield of Cork. Their obſervations, with thoſe of Mr. Maſon, in Donegal, as publiſhed
in the tranſactions of the Royal Society, for the year
1770, and a few others, have authorized me to
make ſome conſiderable changes, in delineating the
form of this iſland; and thereby to give a truer repre-
ſentation, as I apprehend, of the ſhape and ſize of it,
than what any of the printed maps exhibit.

    It is with great regret that I ſeem to caſt any reflec-
tion on the uſeful labours of Dr. Smith; but I think
it my duty to mention, in this place, that although
his authority is very reſpectable in other matters; little
reliance is to made on the latitude and longitude
which he aſſigns to places, from his own obſervations.
For it is evident, from the proceſs * he made uſe of to

    * *Smith's* Ancient and Preſent State of Cork, Vol. I. *Book* I. Ch. 4.

aſcertain

afcertain the longitude of Cape Clear, by obferving an eclipfe of Jupiter's firft Satellite, that he was little acquainted with aftronomy. The * tables which he mentions, gave 9 h. 25 min. 17 fec. P. M. for the time of its immerfion at London, on the 8th of September 1747. He fays, it was obferved by him at 10 h. 4 min. 15 fec. P. M.—and hence he concludes the place of his obfervation, near the Cape, to be 39 min. of time, or 9 deg. 45 min. *weft* from London: whereas, if his time-keeper were truly regulated, and his obfervation accurate, as the tables were probably right, the true conclufion would be, that Cape Clear is fituated 39 min. *eaft* of London. But the Cape being certainly more than 9 h. 15 min. *weft* of London, the apparent time of the eclipfe at Cape Clear, muft have been at leaft 37 min. earlier than at London, or upwards of an hour and a quarter *before* the time at which Dr. Smith fays he obferved it.

To do all that was in my power towards improving and correcting the geography of this kingdom, I employed two fummers in vifiting the different counties, and particularly the remote parts, for which I had not any authentic documents; and in the courfe of thefe tours I collected much information from gentlemen

* Caffini's Tables, rectified by Pound.

of

of knowledge and obfervation, concerning thofe diftricts with which they were well acquainted.

With regard to the *ecclefiaftical* part of this Map,
Sir William Petty's furveys, as mentioned above, were
of the utmoft importance in afcertaining the fituation
of parifhes, and the extent of bifhopricks.

I was, befides, liberally fupplied, from the regiftries
of the feveral diocefes, with every kind of information
that I wanted, and they contained ; for which I am to
acknowledge my obligation to the feveral Bifhops, as
well as for the flattering encouragement with which I
have been honoured by their lordfhips, during the
progrefs of this work.

I have only now to intreat the indulgence of the
public for the errors and inaccuracies which, after all
my pains, will be found in the Map.   With fuch ma-
terials as this country yet affords, many were certainly
unavoidable ; and fome, I fear, are to be imputed to
the infufficiency of my own judgment.   For, in thofe
cafes where certainty was wanting, I have been forced to
recur to reafoning and conjecture : and at the fame
time that I deprecate a hafty cenfure of the work, I
earneftly folicit correction ; and fhall thankfully amend
every fault in it, which thofe, who are better informed, will have the candour to point out.

2   The

The immediate object of the following memoir is
to point out the principal defects of the former Maps
of Ireland, which are amended in mine ; and, in illuf-
tration of the new Map, to give a fhort defcription of
the feveral counties, with refpect to their foil, extent,
population, and commerce ; together with a fummary
account of the ecclefiaftical eftablifhment in each diocefe.

The neceffary limits of fuch a memoir preclude
more than a mere fketch of the prefent ftate of the
kingdom.   But if I fhould have the happinefs of
finding, that this effay is received with indulgence ; I
may perhaps, at a future day, offer to the public a
more full and particular account of Ireland.   In the
part which relates to the Church, I fhould propofe to
trace the hiftory of each diocefe, and to mark the various alterations that have taken place in its eftablifh-
ments, from the earlieft period.   I fhould alfo endeavour to ftate with precifion the circumftances of every
parifh ; fuch as their patronage, extent, impropriations
and glebes ; the valuation in the king's books, the firft-
fruits, crown-rents, procurations, and other charges to
which they may be liable ; with any peculiarities that
relate to them.—In the other part, I would treat the
topography of the country hiftorically ; and not only
                                                      defcribe

defcribe what is interefling at prefent ; but minutely
enquire into the feveral changes that have been made
in the divifions of provinces and diftribution of counties ; in the names of places, the face of the country,
and the territorial property. By confidering the grow-
ing profperity of the kingdom, the gradual but vaft
increafe of its population and opulence, and the con-
nection of each with its efficient caufe ; I would trace
the operation of internal diffentions, and foreign inva-
fions, in former ages: and the happy confequences that
flow from the fettlement of the country, the progrefs of
civilization, and the improvement of arts, manufactures
and commerce, in later times.

At the end of this memoir, I fhall give an explanation of thofe Irifh words, which occur moft frequently
in the names of places ; and fo copious an Index to the
Map, with references to facilitate the finding of any
place, and at the fame time fhew what it is; that it may
be confidered as a topographical repertory to the
kingdom of Ireland.

DESCRIPTION

DESCRIPTION of the MAP.

THOSE parts of the fea coaft, which are bounded by ridges of lofty and abrupt rocks, are fo clearly marked by the graver, that it is unneceffary to do more than mention, that the diftinction is made.

The boundaries of the BISHOPRICKS are expreffed by a *chain* of fmall *pearls*; and where they coincide with the bounds of counties or baronies, the pearls are intermixed with the *round* or *long dots*, by which thofe bounds are refpectively denoted.

But that the *civil* and *ecclefiaftical* divifions may be clearly diftinguifhed at the fame time; the limits of every diocefe may be illuminated by a broad pale colour, while the counties and baronies in each province, are coloured in the ufual manner.

To point out the fituation of the parifhes, I have placed every church in its proper fite; and the eye will at once diftinguifh exifting churches from fuch as are in ruins. To thefe I have added, wherever they remain, as an interefting object to the curious antiquary, thofe fingular buildings, which are peculiar to Ireland, the *round towers*. Thefe towers are all cylindrical, and of ftone; they vary in height from 50 to 140 feet, and from eight to twelve feet in diameter, in the clear. Some are of excellent, and even elegant workmanfhip, and others of very rude mafonry; but all without ftairs. They have commonly four windows, very near the top; and the door, which is elevated ten or twelve feet above the ground, is turned towards the church; on

the

the north-weſt ſide of which they are generally ſituated, though at various diſtances.

But of the age in which they were erected, or of the uſe which was made of them, no *certain* account has yet been collected from Iriſh hiſtory or tradition.

I have appropriated the Roman character excluſively to the names of pariſhes, ſo that when a town or village bears the ſame name as the pariſh in which it lies, it is expreſſed in roman characters; but when they differ, the name of the pariſh is in Roman, and that of the town or village in *Italicks*.

The letters R. V. &c. which follow the name of each pariſh, ſhew whether it be a rectory or vicarage, &c.

A ſingle line under the name of a vicarage denotes that the rectory is a lay impropriation; and a double line, that the tythes of the whole pariſh are impropriate.

The names of all cities, towns, and boroughs, which have the privilege of returning members to parliament, and thoſe only, are expreſſed in capital letters.

I have taken care to repreſent the mountains in a ſuch a manner as might nearly ſhew the ſpace they occupy; I have alſo endeavoured to give an idea of their comparative height, by the varied ſtrengh of ſhading; and the engraver has, except in a very few inſtances, exactly followed my drawing.

The inſertion of the high roads, would have much crowded the map: and as they ſerve only to miſlead, unleſs very accurately deſcribed; I judged it beſt, upon the whole, to mark thoſe only by which the mails are conveyed to the ſeveral poſt towns. The direct poſt roads from Dublin are marked by a *double line*, the croſs poſts by a *ſingle* one, and the poſt towns are diſtinguiſhed by

figures,

figures, which fhew at the fame time how many poft days they have in every week.

It may be fatisfactory to fee with what noted places in England, and even on the continent, the different parts of Ireland agree in latitude. I have therefore marked on the eaftern fcale line of the map, the parallels of fome of the principal towns in England, and of a few cities in Europe and Afia; and on the weftern fcale line, the relative fituation of North-America to this ifland. The fcale lines on the north and fouth, will fhew how much more to the weft Ireland is fituated than any other part of Europe.

C

MEMOIR

TO ILLUSTRATE

A NEW MAP OF IRELAND.

I. CONSTRUCTION OF THE MAP.

THE maps of Jeffereys and Bowles place DUBLIN nearly in the true latitude; but they are very erroneous with respect to its longitude. For the late Dr. Usther found, by the result of a multitude of observations, that the latitude of his observatory in Mecklenburg-street, where he then resided, is 53° 21′ 2″, and the longitude 25 minutes of time, or 6° 15′ 0″ west from Greenwich. Whereas Jeffereys places it in 6° 30′ 0″ and Bowles in 6° 39′ 0″, the longitude in his map varying from 9 to 11 minutes west of the other.

Dr. Longfield assigns to the city of CORK, the latitude of 51° 53′ 54″, and the western longitude of 8° 30′ 0″. But Jeffereys' map places it in latitude 51° 45′ 0″, which is an error of 8′ 54″; and in longitude 8° 37′ 30″; which is only 2° 7′ 30″ west of Dublin, as placed by him: whereas the difference between those two cities is 7′ 30″ greater, according to the observations; the meridian of Cork being 2° 15′ 0″ west of that of Dublin.

At

At Cavan, near the town of Lifford, in the county of Donegal, Mr. Mafon erected a temporary obfervatory, by appointment of the Royal Society, in the year 1769, to obferve the tranfit of Venus; and continued his obfervations there, from April to December. The mean of thefe obfervations, which may be feen in the Philofophical tranfactions for 1770, determines the fituation of that fpot, to latitude 54° 51′ 41″, and to weft longitude 7° 23′ 0″. By the former maps it is placed in lat. 54° 49′ 40″, and in long. 7° 53′ 0″.

But as there can be no doubt of the accuracy of the obfervations which I have now reported; *Dublin, Cork*, and *Cavan*, were laid down in my map, as three fixed points, from which the relative fituation of other places, and the form, and extent of the ifland, might be afcertained with more precifion, than had been done hitherto. There was ftill wanting a pofition in the weft of Ireland, aftronomically determined : and a pupil of Dr. Ufher undertook a journey to Galway exprefsly for the purpofe of obtaining it. By repeated folar, and a few lunar obfervations, he afcertained the latitude of Galway, to be 53° 16′ 0″, which Jeffereys' map makes only 53° 10′ 30″. And though he was prevented by a feries of fhowery and clouded nights, which prevailed the whole time he was able to remain there, from making fuch a number of lunar obfervations, as might fettle the longitude aftronomically : yet as this town lies fo nearly in the fame parallel with Dublin, whofe longitude is accurately fixed ; and as the meafured diftance is by many geometrical methods carefully afcertained, we cannot err confiderably in the determination of its longitude.

The true pofition of *Cavan* being eftablifhed, it enabled me to adjuft the diftances and bearings of the county of Donegal from Mr. Macrea's large furvey, and thereby to determine the fituation of Londonderry, as the map of Donegal neceffarily comprehends that city.

The

The county of TYRONE, which is contiguous to Donegal, and extends eaftward as far as Lough Neagh, was laid down with the like precifion from the furvey of the fame artift. And it is a proof of his accuracy, that I found but a very trifling difference between the fituation of the village of Cookftown in his map, and its true place, according to Dr. Hamilton's obfervations: which from his having had for feveral years a fixed obfervatory there, and having very frequently compared the paffages of the moon's limb reduced to the centre, with the actual obfervations of Dr. Mafkelyne, on the fame days, and omitted few opportunities of obferving the eclipfes of Jupiter's firft fatellite, may be efteemed of incontro-vertible authority.

The map of LOUGH NEAGH and the adjacent country, ferved to connect Tyrone with the counties of Antrim, Down, and Ar-magh, all of which have been furveyed, as was mentioned in the preface. The county of Louth bounds with thofe of Down and Armagh, and a fmall part of the county of Meath intervenes be-tween Louth and Dublin. But here we have the city of *Dub-lin* for another fixed point, which communicates a great degree of exactnefs to the relative pofition of the counties of Meath and Louth, to Kildare, Wicklow, and the Queen's County ; of all which there are modern furveys.

Mr. Scalé's furvey of the courfe of the river *Boyne*, was of great ufe in rectifying the topography of the county of Meath. And by comparing his furvey of the river Brofna, and a plan of the northern line, which had been formerly propofed for a canal from the capital to the river Inny, and fo into the Shannon,—with the meafurement of the roads that lead from Dublin to different towns on the Shannon, we fhall go near to determine the dif-tance of that river from the eaftern coaft, or the difference of longitude between Dublin and Athlone, &c. which I apprehend to be fomewhat lefs than is ufually reprefented. By the maps, *Athlone* is weft of Dublin 1° 42' 0", and the rhomb diftance is 56

2   miles.

miles. But Sir W. Petty makes it only 51 miles : and as the road meafures 59 miles 5 furlongs, and as one mile in eight is the leaft that can be allowed, on an average, for the windings of roads between any two places ; the rectilinear diftance cannot be fo great as 56 miles. In the new map it will be found 52, and the difference of longitude no more than 1° 35′ 0″, which is 7 minutes lefs than in Jeffereys', and in my opinion fo much nearer the truth. I have found it neceffary likewife, to make an alteration in the latitude of Athlone, by placing it in 53° 23′ 30″. To this I was partly induced, by an obfervation of the young gentleman who took the latitude of Galway for me ; but I fhould not have depended on fo few obfervations as he had time to make in paffing through Athlone, had they not been confirmed by the pofition of Athlone, in the furvey of the Shannon, and by Mr. Pelham's map of the county of Clare, which includes a great part of that river. By that map, the difference of latitude between *Galway* and *Loophead*, at the mouth of the Shannon, is 0° 45′ 30″ ; and by the furvey of the Shannon, Athlone is about 53′ 0″ north of the fame cape, a difference of 7′ 30″ : fo, the latitude of Galway being 53° 16′ 0″, that of Athlone muft be 53° 23′ 30″, which is one minute more than in the printed maps.

It is evident that thefe authorities muft equally decide the fituation of Limerick ; which, with refpect to its bearing from Dublin, varies but very little in this map from the old ones, the difference of longitude being the fame in both, and the difference of latitude exceeding the old maps only 32″.

The new map, on the fame authority, reprefents the courfe of the *Shannon*, as trending much lefs to the fouthward, than in thofe of Jeffereys, &c. the difference of latitude between.Limerick and Loophead which they make 13′.30″, being no more than 9′ 0″.

But the true pofition of Cork caufes a great variation in the

bearings

bearings of that City and *Limerick ;* their difference of latitude being only 45′ 6″, and not 49′ 30″, as in the old maps. They also place Limerick 8′ 30″ weft of Cork ; though, in my judgment, there can be no more than *one minute* difference between their meridians. The travelling meafured diftance is 49 miles and a half. But the Rhomb line meafures, on the old maps, 45 miles. On the new map it is but 41, which will be found neareft to the truth, when it is confidered, that the road from Cork to Limerick makes a great deviation from a linear direction, by paffing through Mallow, Buttevant, Charleville, and Kilmallock.

In confequence of M'Kenzie's obfervations, I have placed *Cape Clear* in latitude 51° 19′ 0″, which is 34′ 54″ fouth of Cork. The old maps affign only 33′ 30″ of difference in latitude between them ; but they place the Cape 1° 2′ 30″ weft of Cork. This I conceive to be erroneous, and have therefore followed Sir William Petty's maps, by which the difference appears to be but 55′ 0″, and agrees better with the meafured roads along the fhore.

In tracing the coafts, a particular attention has been paid to M'Kenzie's charts, with refpect to their indentings, fhape, and bearings. But his dimenfions not having been taken by actual meafurement, thofe of Sir William Petty have been moft frequently adopted, where documents of a later date are wanting. For whenever I compared his maps with the modern ones, there appeared fuch a general coincidence of outline and extent, except in a very few inftances, as is furprifing ; when we confider the period at which they were made, and the rapidity of their execution. Sir William Petty's contract with government, is dated December 11, 1654, and the work was finifhed in March 1656.

But in his maps there is no fcale of degrees, whence it is to be prefumed, that his furvey was not adjufted by any aftronomical obfervations. And this will account for the difference in the

general

general bearing of the ifland in the new map, from that which it has in the old ones. By the new projection, the northern part of Ireland inclines more to the eaft, and the fouthern extremity more to the weft, than has been heretofore reprefented. For Jeffereys &c. give but 3° 21′ 30″ of longitude difference between the meridians of *Fair-head* in the north, and of *Mizen-head* in the fouth; while they place them in parallels of latitude 4° 2′ 0″ afunder. Whereas the true difference of longitude is greater by 20′. 0″, and the latitude is 11′ 30″ lefs. But that this new projection muft be the true one, is proved by the obfervations at Cavan, in the county of Donegal, and at Cork; the difference of their meridians being 1° 7′ 0″ which is 22′ 30″ more than in any of the old maps.

In conftructing the fouth-eaft coaft, I have alfo deviated confiderably from my predeceffors, having made the difference of latitude between Wicklow and Hook-tower *three* minutes lefs, and between Hook-tower and Cork near *five* minutes lefs than Jeffereys; which caufes a variation of almoft *eight* minutes in little more than a degree. But it has been fhewn, that there is an error of about *nine* minutes in the old maps, between the parallels of Cork and Dublin. And the point of Hook, thus placed, agrees with Petty's furvey, and with the diftance and bearing from St. David's head in Wales, as laid down in the nautical charts.

The true longitude of Dublin and Cork juftifies the fame charts, in delineating the eaftern coaft, from *Wicklow* to *Carnfore* point, as trending confiderably to the weftward, at leaft *eight* minutes of longitude more than what appears by the old maps; which give only *nine* inftead of *feventeen* minutes of longitude difference between thofe two points.

It is unneceffary to dwell longer on the general conftruction of the map. In the different defcription of the feveral counties, the principal additions, retrenchments, and variations, fhall be noticed

in

in their refpective places. And the following table of the obfervations, by which I was guided, will exhibit to the reader, at one view, how far the fituation of places in the old maps coincides with, or varies from, their true geometrical pofition.

OBSERVATIONS OF LATITUDE.

	Obfervations.			Old Maps.			Error.		
By the Rev. Dr. USSHER.	°	′	″	°	′	″		′	″
Dublin	53	21	2	53	21	0	S.	0	2
Wicklow Pier	52	59	0	52	58	0	S.	1	0
By the Rev. Dr. JAMES ARCHIBALD HAMILTON.									
Armagh	54	20	30	54	20	0	S.	0	30
Cookftown	54	38	20	54	40	0	N.		40
Ardee	53	50	30	53	52	0	N.	1	30
Portarlington	53	9	30	53	11	0	N.	1	30
By the Rev. WILL. HAMILTON.									
Bengore Head	55	15	0	55	15	0		0	0
Ballycaftle	55	12	0	55	14	0	N.	2	0
Londonderry	55	0	0	54	58	0	S.	2	0
By Mr. MASON.									
Cavan, near Lifford, Donegal.	54	51	41	54	49	40	S.	2	1
By Dr. LONGFIELD.									
Cork, City	51	53	54	51	45	0	S.	8	54
By Mr. M'KENZIE.									
Cape Clear	51	19	0	51	11	30	S.	7	30
By a Pupil of Dr. Ufsher.									
Athlone	53	23	30	53	22	30	S.	1	0
Galway	53	16	0	53	10	30	S.	5	30

D

OBSERVATIONS ON LONGITUDE.

	Observations.			Jefferys' Map.			Error.		
	°	′	″	°	′	″		′	″
By Dr. Ussher. Dublin	6	15	0	6	30	0	W.	15	0
By Dr. Hamilton. Cookſtown	6	40	0	7	6	0	W.	26	0
By Mr. Mason. Cavan	7	23	0	7	52	0	W.	29	0
By Dr. Longfield. Cork.............................	8	30	0	8	37	30	W.	7	30

II. OF IRELAND.

THIS iſland, which, next to Britain, is the largeſt in Europe, lies at no great diſtance from the weſtern ſhores of England, and ſtill nearer to the coaſt of Scotland. It is ſeparated from its ſiſter iſland by the Irish Sea, which varies in breadth from fourteen to forty leagues; but is contracted between Scotland and the county of Down to a channel only ſix leagues wide; and farther north, to a ſtill narrower ſtrait, of leſs than four, between the N. E. point of the coaſt of Antrim and the Mull of Kintyre.

This ſea conveys into a few tolerable harbours, on the eaſtern coaſt, the greater part of the ſhipping employed in the intercourſe between theſe kingdoms, and alſo a large proportion of thoſe veſſels which are occupied in foreign commerce.

The principal ports on the eaftern fide of Ireland are, DUBLIN, BELFAST, DUNDALK, DROGHEDA, and WEXFORD. On the northern coaft there are fome bays of confiderable extent; but, except *Lough Foyle*, the bottom of which is the harbour of LONDONDERRY, they contribute little to the general trade of the kingdom. While the advantage of fituation, and the excellence of the harbours of CORK and WATERFORD, in the fouth, have long enabled thofe cities to carry on a very confiderable and daily-improving traffic; augmented by the trade that reforts to the celebrated harbour of KINSALE, and by the fleet of coafters and other fmall craft that crowd the lefs noted ports of YOUGHAL, DUNGARVAN, &c. But the fineft harbours of Ireland are on the weft and fouth-weft. Thofe indented coafts, which prefent innumerable promontories to the fury of the vaft Atlantic, form in their deep receffes fome of the nobleft havens in the world; havens fo fecure and capacious, that in feveral of them the whole navy of Great Britain might ride in perfect fafety. Such are *Kenmare River* and *Bantry Bay, Black-Sod Harbour*, and *Broadhaven*, &c.; which, from the unimproved ftate of the adjacent country, and their diftance from the capital, are as yet but of little ufe.

The weftern coaft is not however deftitute of commerce. The river *Shannon* brings fhips of great burthen to the keys of LI-MERICK, by an intricate navigation of almoft fifty miles from the fea. Much was formerly done at GALWAY, but the bay of SLIGO is now more frequented. There are a few other ports of inferior note, which will be mentioned in their refpective places, as well as the fifheries of *Rutland* and the *Killeries*.

Nature has fortified the moft prominent parts of this coaft, from the north of Donegal to the extreme point of Cork; to ferve as barriers againft the incurfions of the weftern ocean. But neither here, nor in any part of Ireland, are there, as in moft

D 2

other

other countries, long ranges of mountain; if we except one ridge of various heights, and interrupted by the river Blackwater, which extends from near Dungarvan to the county of Kerry. They ftand rather in unconnected groups or maffes, of different magnitude, which are fo difperfed through the ifland, that there are few parts of it in which the profpect is not fomewhere terminated by this fpecies of majeftic fcenery, forming a back-ground feldom more remote than twenty miles. It cannot however be called a mountainous country: though fome counties are hilly, yet many are tolerably level, and others quite flat. From Dublin to the bay of Galway, a vaft plain ftretches itfelf acrofs the kingdom. And in this plain lies the *Bog of Allen*, which was formerly much larger than it is at prefent; a great portion of it having been reclaimed: but it ftill occupies a confiderable tract in the county of Kildare, the King's County, and Rofcommon; and branches off into Meath, Weftmeath, and the Queen's County. This bog, though apparently flat, lies very high, and far above the level of the fea; for the river *Boyne*, the *Little Barrow*, and feveral inferior ftreams, take their rife in it, and purfue their various courfes towards oppofite points of the compafs.

Ireland is indeed extremely well fupplied with water by clear and lively rivers, and innumerable rivulets, by fome large and even magnificent lakes, and by fmall ones without number. If to this beautiful variety of water, hill, dale, and mountain, a fufficient quantity of wood were united, few countries could boaft a greater number of interefting views and picturefque fcenes. But the policy* of more barbarous days, and the demands of luxury in modern times, have applied the axe with fo mercilefs a

---

* Immenfe quantities of wood have been confumed in furnaces for making iron, and in the founderies for cafting metal; of which there have been a great number, and many ftill remain in various parts of the kingdom.

hand,

This Sketch shews the position of the Groups of MOUNTAINS, the extent of the greater Bogs, & the influence they have on the Origin & course of RIVERS.

All the Rivers are shaded as far as they are navigable, & the lowest Bridges are marked.

hand, that of the forests, with which Ireland is said to have been covered, a vestige scarce remains.—The little that is left, at Killarney and other favoured spots, shews what many places have been, and still might be : but as a spirit of planting is diffused all over the kingdom, it may, before very many years elapse, once more become a wooded country.

The principal river of Ireland is the SHANNON, which almost insulates *Clare* and the province of *Connaught* ; and, in a course of 150 miles, expands into six different lakes, several miles long, and from two to six broad. Among the other rivers of note, the FOYLE and the BANN run northward ; the BOYNE, the LIFFEY, and the SLANEY, empty themselves into the Irish sea ; the SUIR, the BARROW, and the NORE, which pour their united streams into the bay of Waterford, with the BLACK-WATER and the LEE, run all to the southward ; the GUIBARRA, the ERNE, the MOY, the MANG, the LANE, and the SHANNON, flow into the Atlantic.

The soil of Ireland varies from the stiffest clay to the lightest sand ; but of the last there is not much to be met with, neither is chalk to be found in any part of it. It is in general much more stony than the soil of England ; and in some large districts the surface appears more than half covered with rocks. Great part of the kingdom lies upon a stratum of rock, at various depths, so that stone quarries abound almost every where : and much of this rock being limestone, it greatly contributes to enrich and improve the land. Marble of great beauty is found in several counties. Mines of coal, iron, lead, and copper, are not unfrequent ; and many of them are worked to great advantage.

The bogs, which supply most of the inland part of the country with fuel, produce on their surface heath, rushes, and coarse grass, with some other aquatic plants, and are generally pasturable in summer, especially on the sides of hills or mountains : and
those

thofe which are in lower fituations become excellent meadows, when thoroughly drained.

However the foil may vary, it is by nature remarkably fertile; and the pafturage is generally thought to be more luxuriant than in England: but in cultivation and good hufbandry the Irifh are ftill much behind their neighbours. Encouraged, however, by the exertions of the Dublin Society, by the advantages which an excellent fyftem of corn-laws prefent to him, and by the example of men of fortune, who all keep large demefnes in their own hands, and many of whom pay the greateft attention to agriculture, the poor hufbandman is daily improving both his practice and his circumftances.

` The climate is rather more variable, and perhaps milder than that of England: the fummers lefs hot, the winters lefs fevere. The air is certainly damper; but this quality is not to be attributed entirely to the bogs which are fcattered all over the kingdom, but chiefly to its infular fituation, and to the great quantity of moift particles that are wafted from the ocean by the wefterly winds, which moft frequently prevail. This moifture, however, is not prejudicial to health, neither is the neighbourhood of bogs unwholefome. The bog waters, far from emitting putrid exhalations, like ftagnant pools and marfhes, are of an antifeptic and ftrongly aftringent quality; as appears from their preferving for ages, and even adding to the durability of the timber, which we find univerfally buried beneath their furfaces; and from their converting to a fort of leather the fkins of men and animals, who have had the misfortune of being loft, and of remaining in them, for any length of time.

Thus does the vicinity of a bog widely differ from thofe apparently fimilar fituations in other countries, which are rendered confeffedly unhealthy by fens or marfhes; but of which there are none in Ireland.

Whether

Whether it be owing to the foil or the climate, certain it is, that in Ireland there are neither *moles*, nor *toads*, nor any kind of *ferpents*; and it is not more than feventy or eighty years fince *frogs*, of which there are now abundance, were firft imported from England. But though the fame experiment has been made with fnakes and vipers, it has happily been unfuccefsful. Wolves were extirpated by Oliver Cromwell. But if this ifland be free from fome noxious, and all venomous creatures, it is, on the other hand, denied one of the fweeteft of the feathered tribe. The nightingale is not to be found there, and when brought over in a cage, but lingers out a miferable exiftence for a fhort time. There are alfo fome other birds, and feveral kinds of fifh, which abound in England, but are unknown in Ireland.

In confidering the *Extent* of this country, I fhall firft obferve, that the greateft difference of *latitude* between the extreme points in the north and fouth is 4° 4′ 0″; the latitude of CAPE CLEAR being 51° 19′ 0″; and that of MALIN HEAD, in the county of Donegal, 55° 23′ 0″. The extremes of *longitude* are 5° 19′ 0″ and 10° 28′ 0″, which give a difference of 5° 9′ 0″ between the moft eafterly part of the county of Down, at BURR ISLAND, and the moft weftern of the BLASKET ISLES, off Dunmore Head, in Kerry.

The greateft *length* of Ireland extends from North-eaft to South-weft. And a line fo drawn between the two moft remote points, FAIR-HEAD and MIZEN-HEAD, would cut the meridian in an angle of 30 degrees, and meafure 241 Irifh miles, which fomewhat exceed 306 of Englifh ftatute meafure. The longeft line that can be ftretched acrofs the kingdom, would meafure 163 Irifh (or 207 Englifh) miles, from EMLAGH-RASH in Mayo, to CARNSORE POINT, in the county of Wexford: and this line would interfect the former in an angle of 75 degrees.

But

But from the STAGS OF CORK HARBOUR to BLOODY FAR-
LAND POINT, in Donegal, is the greateſt length that can be meaſured along a meridian; and it will not exceed 185 Iriſh, or 235¼
Engliſh miles. If the breadth be meaſured in the ſame manner,
nearly on a parallel of latitude, the true breadth of the iſland will
appear to be as follows:

	Iriſh.	Eng.
Between *Tiellen-Head* and *Iſland-Magee* ———	98	124
Between *Emlagh-Raſh* and the mouth of *Strangford Lough* —— —— —— ——	143	182
Between *Slime-Head* and the point of *Hoath*	137	174
Between *Dunmore-Head*, and *Greenore-Point* in Wexford —— —— ——— ——	136	173

But there is not a ſpot in the kingdom fifty miles diſtant from
the ſea; which will not appear ſurpriſing, when we obſerve, that
between the bays of Dublin and Galway, there are but 86 miles,
and no more than 67 between Dundalk and Ballyſhannon.

With reſpect to the ſuperficial contents of this kingdom; not
being able to diſcover any documents of authority in the public
offices, I have made a computation of it, by very carefully mea-
ſuring the area of each county on my map: and after rejecting
all fractions, I have no heſitation in aſſerting, that Ireland contains conſiderably more than 18,750 *ſquare miles*, or ſeveral thouſand *acres* above TWELVE MILLIONS Iriſh meaſure; which is
equal to 30,370 Engliſh miles, or 19,436,000 acres.

In this ſpace there are not fewer than 700,000 * houſes, which
is

---

* By the laſt returns of the officers employed in collecting the hearth-money, to the end of
1790, the number of houſes in the ſeveral counties of Ireland amounted to 677,094. But I
am informed by Mr. Buſhe, one of the commiſſioners of the revenue, who has paid a particular attention to this ſubject, and who obligingly ſupplied me with thoſe returns; that notwithſtanding the ſucceſsful exertions of the board, for ſome years paſt, in improving this
branch of the revenue, by the ſuppreſſion of falſe and incorrect returns, the deficiencies which
ſtill remain cannot be computed at leſs than 23,000.

is at the rate of 37 * houfes to every fquare mile, or 17.7 acres to a houfe, upon an average of the whole kingdom. But there is an aftonifhing difproportion in the population of different counties. In one there are 77, in another but 18 houfes to a mile, as will be found in their refpective defcriptions.

It would much exceed the intended limits of this memoir, were I to examine into the early ftate of Ireland, and to compare the various defcriptions and diffimilar pictures of it, that have been drawn at different periods; I fhall not therefore enter into a difcuffion of its ancient divifions, of the changes that have taken place in them fince it became fubject to the crown of England, or of the different periods, from Henry II. to Charles I. in which the feveral counties were formed into *fhire ground*. Neither fhall I confider the alterations that have been made in the number, the extent, or the union of bifhoprics. It will be fufficient for our prefent purpofe, to mark thofe civil and ecclefiaftical divifions which have been eftablifhed, without alteration, for above a century and an half.

IRELAND is divided,—with refpect to its civil or political diftinctions, into *four* provinces, ULSTER, LEINSTER, CONNAUGHT, and MUNSTER; which are fubdivided into thirty-two counties, and contain 252 baronies, and 2436 parifhes:— and with regard to its church-eftablifhment, into *thirty-two* diocefes, which are united or confolidated under *eighteen* BISHOPS and *four* ARCHBISHOPS.

I fhall firft give a *topographical* fketch of the kingdom, and then proceed to lay before the reader a fhort account of its *ecclefiaftical* ftate.

The province of ULSTER comprifes the *nine* northern counties,—LEINSTER the *twelve* Eaftern,—CONNAUGHT the *five* Weftern,—and MUNSTER the *fix* Southern counties.

---

* Irifh fuperficial meafure is to Englifh as 98 to 61; the number of houfes in Ireland are therefore at the rate of 23 to an Englifh fquare mile, or 27.7 acres to a houfe.

Of

Of thefe Munfter is the largeft, containing 5275 fquare miles. Ulfter contains 5000. Leinfter 4356; and Connaught, which is the fmalleft, 4108.

The Subdivifions of thefe Provinces are as follow:

		Baronies.	Parifhes.	Acres.
ULSTER 9 *Counties.*	*Armagh — contains	5	20	181,450
	Down — —	8	60	348,550
	Antrim — —	8	77	387,200
	Londonderry —	4	31	318,500
	Donegal — —	5	42	679,550
	*Tyrone — —	4	35	463,700
	Fermanagh —	8	18	283,450
	*Cavan — —	7	30	301,000
	*Monaghan —	5	19	179,600
		54	332	3,143,000
Lough Neagh covers — — — — —				58,200
				3,201,200
LEINSTER 12 *Counties.*	Louth — —	4	61	110,750
	Meath — —	12	147	327,900
	Dublin — —	6	107	142,050
	Wicklow —	6	58	311,600
	Wexford —	8	142	342,900
	*Kilkenny —	9	127	300,350
	*Carlow — —	5	50	137,050
	*Kildare — —	10	113	236,750
	*Queen's County —	8	50	235,300
	*King's County —	11	52	282,200
	*Weftmeath —	12	62	231,550
	*Longford —	6	23	134,150
		97	992	2,792,550

* The thirteen *inland* counties are diftinguifhed by an Afterifk.

CON-

				Baronies.	Parishes.	Acres.
CONNAUGHT 5 *Counties.*	Galway	-	-	16	116	989,950
	Mayo	-	-	9	68	790,600
	Sligo	-	-	6	39	247,150
	Leitrim	-	-	5	17	255,950
	*Roscommon	-	-	6	56	346,650
				42	296	2,630,300
MUNSTER 6 *Counties.*	Cork	-	· -	16	269	1,048,800
	Kerry	-	-	8	83	647,650
	Clare	-	-	9	79	476,200
	Limerick		-	9	125	386,750
	*Tipperary		-	10	186	554,950
	Waterford		-	7	74	262,800
				59	816	3,377,150
				252	1436†	12,001,200

ULSTER.

There are 214800 houses returned in this province; which is at the rate of 14.9 acres to a house, or about 43 houses to a square mile.

COUNTY or ARMAGH.

THE length of this county from North to South is 25 miles, the breadth from East to West 15, and the superficial content 181450 acres, or 283 square miles ‡.

† All fractions having been excluded from this calculation, it is very much *under* the full number of acres in Ireland.

‡ In English measure the length is near 32, the breadth almost 20. The area upwards of 454 square miles and 273,786 acres.

It

It is divided into *five* baronies, ONEIL-LAND, ARMAGH, TY-
RANNY, FEWS, and ORIOR, and contains 20 * parishes; of
which 17, comprehending 23 churches, are in the diocese of *Armagh*; and 3 parishes with 3 churches in the bishopric of *Dromore*.

In this county there is very little flat ground, but the gentle
hills which diverfify the face of it are covered with a very rich
foil, except in the South and West parts of *Fews* barony, which
are full of mountains; and in the South of *Orior*, which is occupied by the lofty *Sliebbgullen*. But great part even of this rough
ground is cultivated, and thickly inhabited; Armagh being the
most † populous county in Ireland.

It contains ‡ 21983 houfes, in which there cannot be fewer
than ‖ 120000 inhabitants. This population is at the rate of
8 acres and one fifth to each houfe, or of nearly 78 houfes, and
429 fouls to every fquare mile, which is fomething lefs than one
acre and a half per head.

It is to the great induftry of the people, and to the flourifhing ftate of the linen manufacture which they carry on in all its
branches, that this extraordinary population is to be attributed.

There are feveral good market-towns, and a great number of
villages in this county, the principal are ARMAGH, LURGAN,
BLACKWATER-TOWN, and PORTADOWN.

The city of Armagh, which was very much decayed, has been
renovated, and is become a pretty town, of good fize, and well

* Each parifh- contains 9072 acres, and 6000 fouls, on an average.

† There are more people in proportion to the number of acres, in the county of Dublin,
but that county cannot be taken into the fcale, on account of the capital.

‡ This is the number made up from the returns of the collectors of hearth-money to the
end of 1790, and we may be affured, that whatever it be under, it does not exceed the
truth.

‖ At 5½ fouls to a houfe. In 1075 houfes of every defcription, taken promifcuoufly in
this county, 6988 inhabitants were found; which is upwards of 6 per houfe on an average.
See *Mr. Bufhe's* paper on this fubject in Tranfactions of royal Irifh academy for 1789.

inhabited, through the attention and munificence of the prefent Lord Primate; who has built there a handfome archiepifcopal palace, and a noble houfe for the fchool, which is one of the royal foundations, and extremely well endowed. To thefe his Grace has added a public library for the promotion of fcience. He has alfo erected a complete obfervatory; with a liberal eftablifh-ment for the fupport of an aftronomer; and has fecured the per-manency of his endowments, by feveral acts of parliament obtained for that purpofe.

There is no river of confequence in this county, but it is bounded on the North by *Lough Neagh*, and on the North Weft by the *Blackwater*, which is navigable for fome miles into the Lake. On the Eaft fide, the river *Bann* and the *Newry-canal* af-ford a water-carriage from L. Neagh to the bay of Carlingford. Some good marble is found in this county.

Six members are fent to parliament from Armagh; two for the *county*, two for the *city*, and two for the borough of *Charlemont*.

Obfervations on the Old Maps.

IN the maps of Jeffereys, &c. the barony of *O Neil-land* is mifnamed *Oneland;* a number of villages are marked which do not exift, and the following are omitted, *Loughall*, *Richhill*, *Keady*, *Croffmagline*, *Mohan*, &c. Thofe maps add 2 miles to the length, and 1 to the breadth of the county.

COUNTY of DOWN.

THIS county lies on the Eaft of Armagh, is almoft as thickly inhabited, and nearly double its fize.

The

The length of Down, from North to South is 40 miles; the breadth from Eaſt to Weſt 31, and the area of the county 348550 * acres, or 544 ſquare miles.

It contains *eight* baronies, and the lordſhip of *Newry; viz.* UPPER and LOWER IVEAGH, KINELEARTY, CASTLEREAGH, DUFFERIN, ARDES, LECALE, and MOURNE; in which are 60 † pariſhes; 38 of them, with 33 churches, in the diocefe of *Down*; and 22 with their 22 churches under the biſhop of *Dromore.*

This county is every where irregular in its ſurface, and about the centre, ſwells into a mountainous tract called *Slicbh-Croob.* The barony of *Mourne* is almoſt covered with a large maſs of very high mountains; one of which, *Slicbh-Donard* ‡, is ſaid to be 3150 feet high, above the level of the ſea; but I doubt that it has ever been accurately meaſured, and am inclined to think it cannot ſo much exceed the known height of *Nephin* and *Crow-Patrick* in the Weſt, and of *Mangerton* in the South of Ireland. If we conſider how large a ſpace of this county is covered by theſe very rude mountains, the population of Down will appear very conſiderable, for it contains 36636 houſes, in which muſt dwell at $5\frac{1}{2}$ per houſe 201500 perſons; and this is at the rate of 9.4 acres to a houſe, or 67.34 houſes to a mile.

Moſt parts of this county are fertile, and delightful to the eye, eſpecially about the river *Bann* and the *Lagan.* An irregularity of ground, well watered, abounding in bleach-greens, and full of neat and cleanly habitations, with an orchard to almoſt every cottage, affords a moſt chearful and pleaſing proſpect of the comfort and opulence of the inhabitants, who are chiefly employed in the linen buſineſs. In the rougher parts of the country they breed a

* In Engliſh meaſure 51 miles long, 39½ broad, and the content 874 ſquare miles, or 559905 acres.

† The average of each pariſh would be about 5800 acres, and 3350 ſouls.

‡ On the ſummit of this high mountain is a very large and remarkable Cairn or Carnedh.

great

great number of horfes, with which the fairs of diftant counties are fupplied.

This county derives equal advantages from its maritime fituation and its inland waters. On the North it joins the town and harbour of *Belfaft*, and at DONAGHADEE, packets are eftablifhed for the conveyance of mails and paffengers to *Portpatrick* in Scotland. The ports of STRANGFORD and KILLYLEAGH, upon the *Strangford Lough*, the bays of KILLOGH and DUNDRUM admit veffels of fmall fize. And by means of the bay of Carlingford, and a large canal from thence, the town of NEWRY carries on a very extenfive trade. A canal is continued from Newry to the * *South Bann*, and fo into Lough Neagh. From this lake to Belfaft, another canal has been lately cut at the expence of the Marquis of Belfaft ; and the river *Lagan*, which feparates Down from Antrim, is alfo made navigable.

DOWNPATRICK, the fhire town, is not fmall, but ‡ *Newry* is very large and commercial, and contains more than 10,000 inhabitants. To thefe muft be added HILLSBOROUGH, where the Marquis of Downfhire has erected a moft beautiful church ; DROMORE the refidence of its bifhop, and feveral other market-towns and good villages.

Fourteen members of parliament are deputed from this *County*, *Downpatrick*, the boroughs of *Bangor*, *Hillfborough*, *Killyleagh*, *Newtown*, and *Newry*.

Obfervations on the Old Maps.

THE villages omitted in Jeffereys' map are *Mill-ifles*, *Kircubbin*, *Portaferry*, *Strandtown*, *Crawfurd's Burn*, *Maghera*, *Narrow-*

* The waters of this river are efteemed fuperior to any other, for the purpofe of bleaching.

‡ Part of the town and lordfhip of *Newry* is in the county of Armagh.

water, and *Warren-point*, and many names of imaginary ones are inferted. *Sliebh Crool* is alfo omitted by Jeffereys.

ANTRIM.

THIS alfo is a maritime county, fituated on the North of Down, extending from North to South 44 miles, and from Eaft to Weft 24; and containing 387200 acres, which make about 605 fquare miles *.

It comprizes the *eight* baronies of MASSAREEN, ANTRIM, TOOME, KILCONWAY, DUNLUCE, CARY, GLENARM, and BELFAST, exclufive of the *county of the town* of CARRICK-FERGUS.

Of 77 † parifhes and 44 churches in this county, one parifh, with its church, is in the bifhopric of *Dromore*, the remainder are in the diocefe of *Connor*.

Though the county of Antrim exceeds that of Down in extent by near 30000 acres, yet it falls fhort of it in the number of inhabitants, upwards of 41000. For it contains only 29122 houfes, in which we cannot fuppofe many more than 160000 fouls. This would give, on an average, 13.3 acres to a houfe, or fomething more than 48 houfes, with their 260 inhabitants to a fquare mile. But as there can be but a fcanty population in the mountainous and boggy country along the Eaftern coaft, which, with a large tract of very rough and high hills on the Weft of Belfaft, occupies near a third of the county; the richer and more fertile parts of it are well cultivated and well inhabited, efpe-

---

* The length of this county is 56 miles, the breadth 30½, and the fuperficies 622,059 acres, or 972 fquare miles of Englifh meafure.

† At a medium every parifh would contain fomething more than 5000 acres, and about 2100 fouls.

cially

cially the South of *Maſſareen* and *Belfaſt* baronies, which are in a high ſtate of beauty and improvement. The linen buſineſs gives ſpirit, employment, and wealth to the whole county.

Of the mountains, *Sleniſh* about the middle, and *Knock-Layd* in the North of the county, are the moſt conſiderable.

Antrim is watered by many ſmall rivers, and abundance of brooks and rivulets ; but the broad and rapid *Bann*, by which all the waters of Lough Neagh are diſcharged into the ſea, parts it from Londonderry.

The principal towns are BELFAST, CARRICKFERGUS, AN-TRIM, LISBURN, BALLYMENA and BALLYMONEY. Of theſe the two firſt are ſea ports, and both ſituated on the *Bay of Carrick-fergus*, or as it is now called, the *Lough of Belfaſt*.

By an accurate enumeration made in 1791, Belfaſt contained 3107 houſes and 18320 ſouls. It is with regard to ſize the fifth, and with reſpect to commerce the fourth, if not the third town in the kingdom. There are upwards of ſeven hundred looms in it, employed on cotton, cambrick, ſail-cloth, and linen. Theſe manufactories, with others of glaſs, ſugar, and earthen-ware, the exports of linen and proviſions, and a conſiderable trade with the Weſt Indies, have rapidly increaſed its im-portance.

LISBURN is large and handſome, equally noted for the neat-neſs of its buildings, and the urbanity of its inhabitants.

At CARRICKFERGUS, the aſſizes are held both for the county at large, and for the diſtrict of the town, which has the privilege of a *diſtinct* county. It was once the firſt ſea-port in the north of Ireland, and defended by a ſtrong caſtle. But the port is little fre-quented and the caſtle no longer garriſoned.

LARNE on the eaſt coaſt is but an inconſiderable place, with a poor harbour.

F

At

At BALLYCASTLE, a fmall port on the weft of Fairhead, the fea has entirely wafhed away a mole or pier, which had been erected at a great expence to protect the harbour. There is a good colliery near it, but very much neglected.

Near BALLINTOY on the fame coaft, there are alfo coal mines, which are rendered fomewhat more ufeful.

PORTRUSH is a fmall fifhing town near the mouth of the Bann.

Between thefe two little ports, that celebrated and fublime pile of bafaltick columns, *the Giants Caufeway*, projects into the fea. The ftupendous promontories of *Fairhead* and *Bengore* are in a great meafure compofed of fimilar ftones; which, in a more or lefs perfect ftate, abound in the high cliffs that form this coaft, and in a large circuit of the inland country.

Twelve members of parliament reprefent this *County*, the town of *Carrickfergus*, and the boroughs of *Belfaft*, *Lifburn*, *Antrim*, and *Randalftown*.

Obfervations on the Old Maps.

In them the county is made two miles too fhort, and one mile too broad. Part of the barony of *Maffareen* is thrown into that of *Belfaft*. *Lifburn* is placed on the wrong fide of the river Lagan, and in the county of Down. *Ifland-Magee* is reprefented as an ifland, though a peninfula with an ifthmus of more than a mile broad. The mountains are very ill defcribed, and *Bengore Head*, the moft Northern point of the county, is neither named nor delineated. The villages omitted are, *Bufhmills, Ardmoy, Cullibacky, Newtown-glens, Gracehill, Parkgate, Glynn, Ballinderry*, and *Aghalee*.

LONDON.

LONDONDERRY.

This county, the greater part of which was given by James I. to an incorporated company of London merchants, lies on the weft of Antrim, and extends 32 miles from North to South, and about the fame from Eaft to Weft, meafuring in area 318,500 acres, and 479 fquare miles *.

It comprifes the *Liberties* of the city of LONDONDERRY, and of the town of COLERAINE, with the *four* baronies of TYREKERIN, KENOGHT, COLERAINE, and LOUGHINSHOLIN.

In thefe there are thirty-one † parifhes; five of which, with fix churches, appertain to the diocefe of *Armagh*, and the remainder, with twenty-three churches, to that of *Derry*. The number of houfes in this county is 25007, which amounts on an average to 12.7 acres to a houfe, or 50.3 houfes to a fquare mile; and may contain ‡ 125000 people.

The linen manufacture profpers through every part of this county, which is not much incumbered with mountains. *Benevenagh* in the north, *Sliebhgallan* in the fouth, and *Cairntogher*, which extends into Tyrone, are all that claim our notice. In the laft-mentioned mountain all the rivers of this county have their fource; except the *Bann*, which has been already mentioned, and the *Foyle* which paffes through the liberties of Londonderry, and wafhes the walls of the city. Over this very wide and deep river a wooden bridge, 1068 feet in length, and of fingular and excellent conftruction, was erected in 1791, and

---

* In Englifh meafure 40½ long and broad; 798 fquare miles, and 511,688 acres.

† There are on an average 10,270 acres, and upwards of 4000 fouls to a parifh, in this county.

‡ The population of this county was found to average but 5.06 to a houfe. See Mr. Bufhe's Paper in the Tranf. of the Royal Irifh Academy for 1789.

com-

completed in the fhort fpace of fifteen months, by an American artift named *Lemuel Coxe.*

Four miles below Londonderry the river expands into *Lough Foyle*, a great bay 12 miles long and 7 broad, and land locked on all fides, the entrance not being above half a mile wide, but having only one deep channel in the middle, between fands and fhallows.

LONDONDERRY is a handfome town, containing about 10,000 inhabitants *, whofe principal commerce is with America and the Weft Indies. It is ftill furrounded with walls, and is the county town. Next to this city, in point of note, is COLERAINE upon the *Bann*, about two miles above the mouth of that river. The falmon fifhery near this town, which has been extremely valuable for a long feries of years, begins to decay through mifmanagement. About this town, NEWTOWN-LIMAVADDY, and MAGHERAFELT, the linen bufinefs is very brifk. There are feveral other towns and villages in this county, among which are MAGHERA, DUNGIVEN, CLADY, CLODY, &c. At *Magilligan*, between Benevenagh and the fea, there is the moft productive rabbit warren in the kingdom.

Eight reprefentatives are deputed to the Houfe of Commons, by the *county*, the *city* of *Londonderry*, and the boroughs of *Coleraine* and *Newtown Limavaddy.*

Obfervations on the Old Maps.

The *liberties of* Londonderry are not marked, and *half* of the liberty of Coleraine is omitted ; fo are the mountains *Benevenagh* and *Sliebhgallan*, and the villages of *Crofs*, *Clady*, and *Tobarmore*. In the old maps this county is fmaller than in the new one, by a mile from north to fouth, and three miles from eaft to weft.

* The number of houfes in 1789 was 1642. Tranfact. Royal Irifh Acad. 1789.

TYRONE

TYRONE.

Immediately fouth of Londonderry, the county of TYRONE extends 33 miles from north to fouth, and 43 from eaft to weft. The area meafures 467700 acres, or 724* fquare miles.

This large county is divided into no more than four baronies ; DUNGANNON, STRABANE, OMAGH, and CLOGHER, or *Upper Dungannon*, which contain only 35 parifhes and 38 churches, of which twenty parifhes and twenty churches are in the diocefe of *Armagh* ; eleven parifhes and thirteen churches in that of *Derry* ; and four parifhes, with five churches, in that of *Clogher*. The counties that have been defcribed are more populous than *Tyrone* ; but the population of this county comes neareft to the medium of the whole province ; for it contains 28,704 houfes, which would amount on an average to 39.64 in every fquare mile, and 16.1 acres to each houfe †.

The foil of this county varies exceedingly. Almoft the whole barony of *Dungannon* is rich and fruitful. It abounds with fmall towns and villages : DUNGANNON, STEWARTSTOWN, COAGH, DONAGHY, COOKSTOWN, POMEROY, CALEDON, ORRITOR, AGHNECLOY, BENBURB, MOY, and many others ; in which linen weaving and bleaching are connected with fmall tillage farms. Near *Dungannon* there are good collieries, and a canal from the little village of COAL ISLAND to the *Blackwater*, opens a communication with Lough Neagh and the furrounding country.

The barony of *Strabane* is very rough ; the mountains of *Munterlony* and *Cairntogher* covering a great part of it. *Beffy Bell* and *Mary Gray* are alfo remarkably high. But in all parts

---

* In Englifh meafure 42 from north to fouth, 54½ eaft to weft ; 116¾ fquare miles, 751,387 acres.

† See ULSTER, page 17.

of

of this country cultivation is creeping, and that not flowly, up the fides of all the hills and mountains that are capable of improvement. In this tract are the villages of NEWTOWN-STEWART, ARDSTRAW, GORTIN, DOUGLAS, and the town of STRABANE, finely fituated for trade on the river *Mourne*, which prefently uniting with the river *Finn*, affumes the name of *Foyle*, and becomes navigable to Londonderry.

The greateft part of the barony of *Omagh* is very poor and mountainous; and the town of OMAGH, though the affize town, is inferior to many others in the county.

In the barony of *Clogher* there is more good land: but CLOGHER, FINTONA, and AUGHER, are very fmall places, notwithftanding the laft is a borough, and the firft a *city*, fince it is the fee of a bifhop, and fends members to parliament.

The *Blackwater* rifes in the fouth of this county; but the fine and rapid rivers which water the heart of it, the *Cameron*, the *Po*, the *Moyle*, and many others, all fall into the *Mourne*.

The towns of *Strabane* and *Dungannon*, the borough of *Augher*, and the city of *Clogher*, with the county, return *ten* members to parliament.

Obfervations on the Old Maps.

The county is reprefented a mile too large in every dimenfion and the following villages are omitted: *Donymanagh, Douglas, Magheracrigan, Gortin, Drumquin, Bellnahatty, Newtown-Saville, Coagh, Orritor, Brockagh, Moy, Eglifh, Dian,* and *Coal Ifland.*

LOUGH NEAGH.

This great body of water, which wafhes the fhores of the five counties that have been defcribed, muft not be omitted. It is

fifteen

fifteen miles long, *seven* broad, and covers 58200 acres. The old maps, which make it larger, are as incorrect in the fize as the fhape of the lake, according to Mr. Lendrick's Survey.

The river *Bann* is the only outlet for feven rivers and innumerable ftreams, that pour their tributary waters into this great inland fea; which, though by far the largeft, is by no means the moft beautiful of the Irifh lakes.

The fhores are moftly formed either by an inanimate ftrand, or marfhy borders liable to frequent floods; and are of courfe deficient in thofe varied banks and bold promontories, without which fuch extenfive fheets of water cannot have a picturefque effect, unlefs when the uniformity is broken by frequent iflands of different fize and character:—but there are only two in this lake; a very fmall one near the mouth of the Blackwater, and *Ram Ifland*, within a fhort diftance of the Antrim fhore, remarkable only for an ancient round tower. The views are more pleafing in *Lough Beg*, a fmall lake, into which thefe waters again expand, after a courfe of about a mile through a very contracted channel. The form of *Lough Beg*, its iflands, fome wooded points of land, with intervening lawns and rocks, the magnificent rotunda at Ballyfcullen, and the beautiful lightnefs of Toome-bridge, produce the moft happy effect.

It would be unpardonable to omit, that *Lough Neagh* has been long celebrated for a petrifying quality, which the water, or rather (it is faid) the foil, poffeffes on fome parts of the Antrim fhore.

DONEGAL *.

This county, the largeft in Ulfter, extends on the weft of Tyrone and Londonderry 57 miles from north to fouth, and

* It has been formerly called alfo TYRCONNELL.

40 from

40 from eaſt to weſt. It contains 679550 acres, or 1061 ſquare
miles *, and is divided into five baronies—INISHOWEN, KILMA-
CRENAN, RAPHOE, BOYLAGH and BANNOGH, and TYRHUGH.
The number of pariſhes † in this county is forty-two ; thirty of
which, with thirty-two churches, are in the diocefe of *Raphoe* ;
eleven pariſhes and thirteen churches in that of *Derry* ; and one
pariſh, with its church, in the biſhopric of *Clogher*. It contains
only 23,521 houſes, in which we cannot eſtimate ‡ more than
140,000 inhabitants ; a population which is inferior to that of
twenty-nine counties of the thirty-two. For it averages but one
houſe to 28.8 acres, or 22.17 houſes in a ſquare mile.

But it muſt be confidered, that *Donegal* is a very rugged country, in many places rendered leſs habitable by bogs, and almoſt
every where rough with mountains. It is not however deſtitute
of good land in the vales between thefe rocky maſſes, and along
the banks of many rivers.

The chief of thefe are, the *Finn*, which, riſing in a lake,
croſſes the county from weſt to eaſt ; the *Dale*, navigable by
boats for a few miles, from the river Foyle to the village of
BALLINDRAIT ; the *Erne*, which runs from *Lough Erne*, and
falls into the fea below Balliſhannon ; and the *Guibarra*, a river
of extraordinary breadth and depth for the ſhortnefs of its
courfe, which extends fcarce twenty miles from its fource to the
fea.

This county is rich in harbours—*Lough Foyle* has been already
noticed. *Lough Swilly* is a prodigious fine harbour, twenty miles
long, from two to four broad, and deep enough for the largeſt

* In Engliſh meafure feventy-two miles long, fifty-one broad ; 1704 ſquare miles,
1,091,736 acres.

† There are on an average 16179 acres, and about 3350 fouls to each pariſh.

‡ The population of Donegal is ſtated, by the return made to Mr. Buſhe, to be at the
rate of 7.35 to a houfe. I have computed them at fix. See Mem. Royal Iriſh Academy
for 1789.

man

man of war. The bay of *Strabragy* in Inifhowen, *Mulroy Bay*
and *Sheep Haven*, in Kilmacrenan; the *Guidore*, the *Guibara*;
Donegal Bay, with *Killibegs* Harbour and many other fmall ones
branching off from it: and the road or harbour at the *Roffes* are
all excellent and fafe retreats for veffels.

To this county belong *Tory Ifland*, about fix miles from the
north-weft point of it, and *Arranmore*, with a clufter of fmaller
iflands, near the coaft of that part of Boylagh and Bannagh,
which is called the *Roffes*. In one of thefe iflands, named *Rutland*, in compliment to the late duke of Rutland, a fmall town has
been lately built by the public-fpirited exertions of the Rt. Hon.
Mr. Conyngham, for the purpofe of promoting the herring
fifhery; to which this fituation feemed peculiarly adapted.

Of the many lakes in this county I fhall only mention *Lough
Derg*, which is fituated in the midft of mountains, in the barony
of *Tyrhugh*, and has been for ages celebrated, on account of a
fmall ifland containing a cell called Saint Patrick's Purgatory;
to which the fuperftitious devotion of the times drew many a
pilgrim in former ages.

The principal town in Donegal is BALLYSHANNON, which
has the advantage of a falmon fifhery in the river *Erne*.   LIF-
FORD the county town, is fcarce a mile from *Strabane*, and but
very fmall.   LETTERKENNY is happily fituated at the bottom
of *Lough Swilly*, but derives no great advantage from that circumftance.   DONEGAL is of little note, notwithftanding the fine bay
on which it ftands.   RAPHOE is the fee and refidence of a bifhop, but otherwife very infignificant.

The town of KILLYBEGS, and a good many tolerable villages,
might be added to this lift; and it muft not be overlooked, that
in this county alfo, the linen manufacture is vigoroufly attended
to.

Twelve members fit in parliament for the county of *Donegal,* and the boroughs of *Ballyſhannon, Donegal, Killylegs, Lifford,* and *St. Johnſtown.*

Obſervations on the Old Maps.

The coaſt is in many places erroneouſly delineated, particu-larly in Iniſhowen. That barony is repreſented as a flat country, and cut in two by a river from *Lough Foyle* to *Strabragy Bay.* *Malin Head* is omitted. *Mulroy Bay,* and the adjacent country, are quite miſrepreſented, and ſo are the *Guibara* and *Guidore* rivers. Tory Iſland is removed at too great a diſtance from the coaſt: the iſle of *Arran* is ſtill more miſplaced, and the ſmaller iſlands which accompany it, and form the harbour of the Roſſes, are omitted. . Neither do they expreſs the villages of *Carrigart, Churchbill, Carrigans, Pluck, Killygordon, Carne, Carrickmaquigly, Ardra, Kilmacreda, Duncanely,* &c. *Rutland* they could not have marked. Of many miſnomers I ſhall only mention that *Horn-head* is written *Horehead,* and *Rackibirn* Iſland called *Raghlin.* The old maps add alſo a mile more to the breadth of the county than the ſurvey allows of.

FERMANAGH.

The greateſt length of Fermanagh, which adjoins to the South of Donegal, is 34 miles, and the greateſt breadth 26 ; the area is 283,400 acres, or 448 ſquare miles : but if we make allowance for the ſpace that is covered by the waters of Lough Erne, which is at leaſt 47,400 acres, we muſt reckon only 236,000 acres of the habitable ground, or perhaps 238,000, by taking the iſlands of the lake into the number *.

* Length 43 Engliſh miles, breadth 33 ditto; 719 ſquare ditto, or 455,298 Engliſh acres—Lough Erne 76,311 ditto—Remainder 378,987 ditto.

This

This great lake completely fevers in two the county. Of its eight baronies, Lurge, Tyreskennedy, Magherastephana, Clonkelly, and Coole, are on the east, and Magheraboy, Clonawly, and Knockninny, on the west of Lough Erne.

These are divided into no more than *eighteen* * parishes : 15 of which, containing 23 churches, are in the diocefe of *Clogher* ; and *three* parishes, with as many churches, are under the fee of *Kilmore*.

This county is but thinly peopled ; the number of houfes in it being only 11,969, which, excluding Lough Erne, is at the rate of 19.9 acres to a houfe, or 32.43 houfes in a square mile. I am however inclined to eflimate the number of fouls † at 71,800.

The furface of Fermanagh is very uneven ; the borders of Tyrone and of Cavan, on the weft of the lake, and efpecially of Leitrim, are extremely mountainous, and the whole county full of hills ; many of them high, rough, and boggy : but even in this rude ftate, thefe hills and mountains afford a coarfe pafture to large herds of young cattle ; and that moft of them are capable of great improvement, and of being brought into tillage by proper management is proved, by the fuccefs which has already attended the exertions of induftry.

Lough Erne confifts properly of two lakes, connected by a broad winding channel of about fix miles. The upper lake is *nine* miles long, and from *one and a half* to *five* wide ; the lower lake extends in length about ten miles, and increafes in breadth from

---

* The average in every parifh is 13,220 acres, and about 4000 inhabitants.

† This is at the rate of *fix* to a houfe, becaufe it appears by Mr. Bufhe's table, that the population of Fermanagh, as far as it was examined into, was found to be at the rate of 7.38 to a houfe. *Tranf. of Royal Irifh Academy*, 1789.

two

two to *eight.* Both thefe lakes are full of iflands *, fome of which
are large and inhabited, many of them well wooded, and the whole
fo difpofed, and accompanied by fuch a diverfity of coaft, as to
form a vaft number of rich and interefting profpects. This lough
receives the *Erne* and feveral other rivers, and difcharges itfelf at
the north-weft end by a rapid current of about feven miles ;
which, after falling over many ledges of obftructing rocks, pre-
cipitates its waters down a grand cataract into the fea at Balli-
fhannon.

There are two other lakes of confiderable length which lie between *Fermanagh* and *Leitrim*; *Lough Melvin* and *Lough Mac-
nean.*

ENNISKILLEN is the only town of note in Fermanagh ; it is
built in an ifland, formed by the river which unites the two
lakes, and is the fingle pafs of communication between the parts
of the county which thefe waters feparate. It has a fchool of
royal foundation ; the endowment of which is, by the great rife
and improvement of lands in this county, become very confiderable. There are befides a few, and but a few, fmall market
towns and good villages: KESH, LISNASKEA, MAGUIRES-
BRIDGE, NEWTOWN-BUTLER, &c. in the eaft; BELLEEK and
GARRISON, &c. in the weftern divifion. On Lord Ennifkillen's
eftate, weft of Lough Erne, there are quarries of a grey or
brown and white marble, beautifully veined, and of a very fine
grain.

The linen manufacture, and the rearing of black cattle, are the
principal fources of wealth to the inhabitants of Fermanagh.

The *county*, and the town of *Ennifkillen*, fend four reprefentatives to the Houfe of Commons.

---

* It is faid, there are between 300 and 400 iflands in this lake. In the ifland of
Devenifh ftands a round tower, the moft elegant of any in the kingdom for its ftyle of
architecture.

Obferva-

Obfervations on the Old Maps.

THEY make Fermanagh about 4 miles too long. I find in them many names of villages which I cannot hear of; but I fee no omiffions except *Kefh* and *Coltrain*. Lough *Maenean* is mifcalled Lough *Cane*.

CAVAN.

SOUTH of Fermanagh lies CAVAN. Its length from E. to W. is 40 miles, and the breadth from N. to S. 22. It contains 301,000 acres, and 470 fquare miles *, and is divided into feven baronies, TULLAGHAGH, LOUGHTEE, TULLAG-HARVEY, CLONCHEE, CASTLERAGHAN, CLONMOGHAN, TULLAGHONOHO; and into 30 parifhes †, *one* of which, and *one* church, are in the diocefe of *Meath*, 3 parifhes and 3 churches in the bifhoprick of *Ardagh*, and the remaining 26 with 24 churches in *Kilmore*.

The number of houfes in this county is 16,314, which may contain 81,570 perfons, and the proportion of population to ex-tent, is 18.4 acres to a houfe, or 34.71 houfes to a fquare mile.

The foil is chiefly a ftiff, wet clay, which produces naturally a coarfe rufhy pafture; but in fome places it has been much a-mended by cultivation. The furface of the county is fo remark-ably uneven, that a level fpot is rare to be met with: a great part of it is open, bleak, and dreary; but from *Cavan* to Lough *Erne* is extremely well wooded and picturefque. Though many of thefe hills are high and barren, yet none merit the appellation of

* Length 51 miles, breadth 28, area 755 fquare miles, or 483,573 acres, Englifh mea-fure.

† Every parifh would contain on an average 10,033 acres, and 2700 fouls.

mountains,

mountains, except *Brucehill* in the Southern extremity, the lofty *Sliebh Ruffel*, which lies partly in Fermanagh, and the mountains of *Ballynageeragh*, which block up the North West angle of the county.

At the foot of these hills are a great number of small lakes, and some of larger size; as Lough *Ramor*, Lough *Sheelin*, and Lough *Gawnah*, on the borders of *Meath*, *Westmeath* and *Longford*: but the most remarkable is *Lough Oughter*, not far from the town of *Cavan*, The irregularity of its form, the large and beautiful* islands it contains, and the many deep recesses that wind between high banks and overhanging woods, produce a rich variety of interesting and romantic scenery. Through this water flows the river *Erne*, which rises on the borders of *Longford* and *Leitrim*, and on its way to the Lough receives many smaller rivers.

CAVAN the county-town is neither large nor commercial, but COOTEHILL has the advantage of a well frequented linen market, in which great sums of money are weekly circulated. BELTURBET is of little consequence; neither is there much to be said of KILLESHANDRA, BALLYHAYS, BAILYBOROUGH, KING'S-COURT, and some other small towns and villages. But SWANLINBAR is justly celebrated, and much resorted to in the summer, on account of its medicinal sulphureous springs.

The *county*, and the towns of *Cavan* and *Belturbet*, are represented in parliament by *six* members.

Observations on the Old Maps.

IN them the length of the county exceeds its just dimensions, by about *five* miles; and the breadth by at least *two*; *Lough*

---

* In one very small bare island, stand the ruins of a castle in which the good bishop *Bedel* was confined by the insurgents in the last century.

Oughter,

Oughter, and the courfe of the *Erne*, are ill defcribed, and a large lake is omitted in the barony of *Tullagha*. *Sliebh Ruffel* is not accurately placed, nor are the fmaller mountains marked in the baronies of Loughtee and Caftleraghan. .

MONAGHAN.

NORTH Eaft of Cavan we enter the county of MONAGHAN, which extends 30 miles from N. to S., and 19 from E. to W.; and forms an area of 179,600 acres, or 280 fquare miles *.

It contains five baronies, TROUGH, MONAGHAN, DARTREE, CREMOURNE, and DONAGHMOYNE; and † 19 parifhes, in which are 19 churches, all in the diocefe of *Clogher*.

This is, next to Armagh, the moft populous county in Ireland, for 76.86 houfes to a fquare mile, and 8.3 acres to a houfe, are the average of the 21,523 houfes in Monaghan, which muft contain about 118,000 fouls ‡, at $5\frac{1}{2}$ per houfe.

The foil of this county is in many places moift, but in general deep and fertile. It is not lefs hilly, but much more diverfified and fheltered by trees than the neighbouring county of Cavan. Neither is it deficient in lakes and rivers. Amongft the lakes thofe of *Kilcrow* near Cootehill, and of *Barrac* at Caftleblayney, deferve notice for their extent and beauty. Of the rivers none are confiderable, but many very pleafant.

The mountains of *Sliebh-Baught* extend from Tyrone a good way into the barony of *Trough*. There are no others in this county which deferve the name of mountain. But there are many rocky hills in the barony of *Magherofs*.

---

* Thefe dimenfions in Englifh meafure, would be,—length 38 miles, breadth 24, area 450, or 288,500 acres.

† About 9450 acres and 5660 fouls to a parifh, on an average.

‡ By the returns made to Mr. Bufhe, the houfes in this county fhould contain each on an average 6.5 fouls. *Tranf. Roy. Ir. Acad.* 1789.

MONAGHAN, the affize town, is not contemptible; CLONES and CARRICKMACROSS are also pretty large, and in a ftate of improvement. To which muft be added BALLIBAY, CASTLE-BLAYNEY, GLASLOUGH, CASTLESHANE, and fome other thriving villages.

The linen manufacture fucceeds admirably, efpecially in the Northern and Weftern parts of this county.

Four members reprefent the *county* and the borough of *Monaghan* in parliament.

Obfervations on the Old Maps.

IN Jeffereys' map the mountains of *Sliebb-Baugh*, and the villages of *Emy-Vale*, *Rockcorrey*, *Drumfwords*, and *Smithborough* are omitted, and *one* mile is taken from the length, while *two* are added to the breadth of the county.

PROVINCE of LEINSTER.

THIS province comprifes *twelve* counties, and they contain 2,792,450* acres, 992 parifhes, and 181,948 houfes; which is at the rate of 15.3 acres to houfe, or 41.7 houfes to a fquare mile.

LOUTH.

THIS is the fmalleft county in Ireland; being only † 21 miles long from N. to S. and 14 broad from E. to W. and containing

---

* Englifh acres 4,460,657.

† In Englifh meafure 26¼ miles long, not quite 18 broad; contains about 278 fquare miles, or 177,926 acres.

no more than 173 fquare miles, or 110,750 acres. It has Armagh on the North and Monaghan on the North-Eaft; and is divided into *four* baronies, DUNDALK, LOUTH, ARDEE, and FERRARD, befide the *county of the town of* DROGHEDA.

Small as this county is, it contains 61 * parifhes in the diocefe of *Armagh*, on which there are 20 churches; and part of two parifhes in the diocefe of *Clogher*.

Louth is the moft populous county in *Leinfter*, the number of houfes in it being 11545, which number, on an average, 66.73 in a fquare mile, and 9.6 acres to a houfe, and contain about 57750 fouls.

There are fome poor hills in the neighbourhood of *Collon*, and a tract of high mountain between *Dundalk* and *Carlingford*: but, thefe rough grounds excepted, the foil of this county is rich and fertile, and chiefly employed in tillage.

Four fmall rivers crofs the county of LOUTH from weft to eaft, and the *Boyne* forms its fouthern boundary.

The bay of CARLINGFORD is a fine haven, with twenty fathom water, but of more advantage to *Newry* than to the town whofe name it bears; which has dwindled into great infignificance, and is remarkable only for the excellence of its *oyfters*. DUNDALK, on the contrary, without the advantage of fo good a harbour, has much trade, and is in a very improving ftate. A cambrick manufactory, which was eftablifhed there fome years ago, has not indeed fucceeded; but thofe of muflin, damafk, and other kinds of linen, are all very flourifhing.

DROGHEDA, which has the privilege of a diftinct county, and its own affizes, is a large well-built town on both fides of the river *Boyne*, and increafes daily in wealth, commerce, and the number of its inhabitants, which amount to more than *ten thoufand* †. This port fupplies the neighbouring country, for many

---

* Thefe parifhes would contain each, on an average, 1788 acres, and 931 fouls.

† The number of houfes in 1789 were 1731. See Mr. Bufhe's paper *ut fupra*.

H   miles

miles round, with Englifh coals and other heavy goods ; and exports very confiderable quantities of corn, the produce of the adjacent and of feveral of the inland counties.   Here is alfo a celebrated fchool, with a very good endowment.

DUNLEER and ARDEE are fmall towns—CASTLE BELLINGHAM is a remarkably well-built and pretty village, noted for the beft malt liquor in Ireland.—COLLON, though not large, is very neat and thriving. By the judicious encouragement of the Rt. Hon. Mr. Fofter, Speaker of the Houfe of Commons, a confiderable thread manufactory, and above a hundred muflin looms, have been added to other branches of the weaving bufinefs. Through all this county the manufacture of brown linens, and a diligent application to agriculture, divide the induftry and attention of the people.

There are a greater number in *Louth*, than in any other part of the kingdom, of thofe high artificial mounts, the fortreffes of early ages, which the Irifh call *Raths*, and attribute to the Danes.

Twelve members of Parliament are elected from this fmall county, if we confider Drogheda as a part of it ; namely, for the *county* of *Louth*, for *Drogheda*, *Carlingford*, *Dundalk*, *Ardee*, and *Dunleer*.

Obfervations on the Old Maps.

In the maps of Jeffereys', &c. this county is reprefented two miles longer, and three miles broader, than in Mr. Taylor's furvey; and the village of *Lurgan Green* is omitted, where there is a fine and extenfive ftrand, covered with a profufion of cockles, which afford a profitable employment, and a wholefome article of diet to the inhabitants.

MEATH.

MEATH.

The county of MEATH is fituated on the fouth-eaft of Louth, and extends from N. to S.* 29 miles, from E. to W. 35, and includes an area of 327,900 acres, or 512 fquare miles.

There are *twelve* baronies in this county—SLANE MORGAL-LION, KELLS, *Half* FOWRE, LUNE, NAVAN, DULEEK, SKRYNE, RATOATH, DUNBOYNE, DEECE, and MOYFEN-RATH; which contain 147 † parifhes, with 44 churches, all in the bifhopric of *Meath*; half a parifh in the diocefe of *Kilmore*, and a fmall part of one in *Armagh*.

The population of this county may be calculated at about 112,400 fouls, the number of houfes amounting to 22,468; which gives 43.88 houfes to a fquare mile, and allows 14.6 acres to a houfe. The foil of Meath is various, but generally rich, and incumbered with very little wafte land; though there are fome coarfe hills about the middle of *Slane* barony, and in the northern part of *Kells* and Half *Fowre*. The bogs are neither numerous nor extenfive in this county; and wherever this is the cafe in Ireland, the inhabitants fuffer from the fcarcity of turf or peat, an article fo effential to their comfort and to the profperity of their manufactures, in a country where timber is very fcarce, and fo few coal mines have as yet been difcovered.

Much coarfe linen is made in this county, but its principal fources of wealth are derived from the flocks and herds that are fattened, and the abundance of corn that is raifed, on its fruitful plains.

* From north to fouth 36, from eaft to weft 44½; area, 526,790 Englifh acres, or 822 fquare Englifh miles.

† The average of each parifh would be 2215 acres, and 760 perfons.

The

The pleasant *Boyne*, as Spenser justly calls it, passes through the heart of the county. The *Blackwater*, which falls into it at Navan, the *Borora*, the *Nanywater*, and many smaller rivers, contribute to its fertility and ornament.

The county town is TRIM, a very ancient place on the *Boyne*, but without manufactures or commerce. *Navan*, on the same river, beautifully situated but very ill-built, is an opulent town, and contains about 4000 inhabitants, most of them industriously occupied in different branches of trade. The commercial interests of this county, and especially of this part of it, will be much improved, when the *Boyne* is made completely navigable from Drogheda to Navan (a work which proceeds with great vigour), and when the projected canals from thence to *Kells*, and to *Trim*, open a new and easy communication with the sea.

KELLS is a good thriving town on the Blackwater—ATHBOY has little trade.—SLANE is a neat village, with some very good houses.—DULEEK and RATOATH are very small and insignificant—DUNSHAGHLIN and TARAH, &c. are among the small villages, which are scarcely worth naming in such a contracted account.

At *New-grange*, near Slane, is a most remarkable Mount, Barrow, or Rath, with a curious chamber in the centre, constructed of rude stones, and accessible only by a long passage, very low and narrow *.

The poor villages of *Ratoath* and *Duleek* have their representatives in the senate, as well as *Navan*, *Trim*, *Kells*, *Athboy*, and the *County*. They return, in all, *fourteen* members to parliament.

* This Barrow is minutely described by Governor Pownal in the *Archeologia*, Vol. II.

Observa-

Obſervations on the Old Maps.

They make the length of the county from eaſt to weſt too great by *two* miles. Many places, which are not villages, are repreſented as ſuch, and ſome inſignificant villages have the appearance of conſiderable towns.

DUBLIN.

The length of this county, which lies on the S. E. between Meath and the ſea, is from N. to S.* 24 miles, and from E. to W. 15. It contains 142050 acres, which make about 221 ſquare miles. It compriſes *ſix* baronies, excluſive of the city and liberties of *Dublin*. On the north ſide of the river *Liffey*, which, running eaſtward, divides both the county and the city, are the baronies of BALRUDDERY, NETHERCROSS, COOLOCK, and CASTLEKNOCK; on the ſouth ſide, thoſe of NEWCASTLE and *Half* RATHDOWN; which ſix baronies, with the city, contain 107 † pariſhes, and 58 churches.

In ſtating the population, we muſt diſcriminate between the county and the city—Taken together, they comprehended, at the end of 1790, 25510 houſes, which gives on an average 5.5 acres to a houſe, and 115.42 houſes to every ſquare mile. But if we ſuppoſe the city to cover about five ſquare miles (3200 acres), and, to avoid fractions, deduct 3050 acres from the groſs content of the

* Theſe dimenſions are in Engliſh meaſure—Length 30½ miles; breadth 19 miles; area 228,211 acres, or 355 ſquare miles.

† Twenty of theſe pariſhes are in the city—The 87 in the county would contain, at a medium, about 1600 acres, and 620 inhabitants each.

county; and if we reckon 10760* houfes in the county, which may contain about 54,000 fouls, the average will be 12.91 acres to a houfe, and 49.86 houfes to a fquare mile; a population which is rather thin for the diftrict that furrounds the capital, being inferior to five of the more northern counties.

This county, indeed, is not to be claffed among the moft fruitful, or the beft cultivated; and towards the borders of *Wicklow* it affumes the mountainous and rocky character of that county. The remainder is flat and uninterefting, except in the neighbourhood of the fea coaft, which being broken into bays and creeks, affords many picturefque and pleafing profpects.

It is impoffible to give even an adequate fketch of the city of DUBLIN in this fhort tract: but it would be unpardonable to omit fome of its principal features. It extends above two miles in every direction, and is rapidly increafing in fize, opulence and beauty. The river *Liffey*, which paffes through its centre, is croffed by five bridges, and a fixth of great elegance is nearly finifhed. The harbour is defended from the inclemency of the winds and waves by a ftrong wall or mole, now almoft completed, extending nearly four miles in length, and terminating with a light-houfe, which is erected about a mile from the eaftern and adjacent to the fouthern Bar. Correfponding to this handfome and ufeful pharos, there ftands on the promontory of *Hoath*, which forms the north fide of the bay, another very complete light-houfe. The vaft number of country houfes and neat villages which cover the fhores of the bay, the varied and undulating fummits of the Wicklow mountains in the fouthern back-ground, and the prof-

* The number of houfes in the county of Dublin, exclufive of the city, in 1788, was 10759, and in the city 14327; and notwithftanding the prodigious increafe of buildings in the laft three years, it may be thought too great an augmentation to flate them now at 16000 inhabited houfes, which, at nine to a houfe, will contain 144000 fouls.

pect

pect of the city at the weft end, compofe one of the grandeft
fcenes that can be imagined.

In this large city there are but 20 parifhes, and 18 parifli
churches, fome chapels of eafe, the cathedral of St. Patrick, and
the collegiate Chrift's church.

Of the many public edifices that adorn it, the moft remarkable
are,—the *Caftle*, which ftands nearly in the middle of the town,
and is the refidence of the viceroy, the *Parliament-houfe*, the
Univerfity, the *Royal Exchange*, and the *Cuftom-houfe*; all buildings of great magnificence. The courts of juftice are almoft
finifhed in the fame ftyle—The *Linen Hall*, the *Barracks*, *Hofpitals* for invalids—for the children of failors and of foldiers—for
the education of youth, and for the reception of the aged and in-
firm of various defcriptions—are too numerous to mention feparately. The ROYAL IRISH ACADEMY was inftituted in 1786—
The DUBLIN SOCIETY, for the improvement of agriculture
and ufeful arts, which was incorporated fo early as in 1749, has
truly anfwered the end of its inftitution, by promoting objects of
the utmoft importance to Ireland.

Two canals are begun on the oppofite fides of the river, with
which they are immediately to communicate. The fouthern navigation extends upwards of 40 miles, to the river Barrow, which
is navigable; and a branch of this canal is carrying on in a
weftern direction towards the Shannon. The northern canal has
alfo for its object, to communicate with that great river in the
county of Longford, and, by a collateral cut to unite with the
Boyne navigation.

To the weft end of the town adjoins the Phœnix Park, a royal
demefne of great beauty, which extends above two miles in
length, and a mile and a half in breadth, on the N. fide of the
Liffey.

The corporation of this city confifts of a Lord Mayor, 24
alder-

aldermen, and a Common-Council, &c. It is almoft needlefs to fay, that moft branches of commerce and every kind of trade are exercifed in Dublin, and that the city is daily advancing in wealth and induftry.

Within a few miles of the metropolis are many pleafant vil- lages—CLONTARF, GLASNEVIN, FINGLASS, CHAPELIZOD, LEIXLIP, LUCAN, whofe fulphureous waters occafion a great refort of company in the fummer months.—RATHFARNHAM, MILTOWN, RINGSEND, the BLACK ROCK, DUNLEARY, and many others. The two laft named, and *Clontarf*, are crowded in the feafon for fea-bathing, and are furrounded with numerous villas of the nobility and citizens, whom the beauties of Dublin Bay attract towards the coaft.

There are alfo in this county the fmall towns or villages of TALLAGH, NEWCASTLE, SWORDS, LUSK, BALRUDDERY, RUSH, SKERRIES and BALBRIGGEN.

Rufh is a large village on the coaft, from whence and from *Skerries*, whofe fmall harbour is rendered fafe by a pier, Dub- lin is principally fupplied with fifh. *Balbriggen* has alfo a fmall fafe harbour, and a flourifhing cotton manufactory.

The *county*, the *city*, and the *univerfity* of Dublin, with the boroughs of *Newcaftle* and *Swords*, return *ten* members to parlia- ment.

Obfervations on the Old Maps.

They delineate the county *one* mile too long, and omit the villages of *St. Margaret*, *Killfologhan*, *Glafnevin*, *Clontarf*, *Ball- doyle*, *Howth*, *Crumlin*, *Clondalkin*, *Bullock*, *Dunleary*, and the *Black Rock*; and in Jeffereys' map *Balbriggen* and *Skerries* are alfo omitted.

WICKLOW.

WICKLOW.

THE extent of this county, which lies immediately South of Dublin, is from N. to S. * 32 miles, from E. to W. 26; and the fuperficial contents are 311,600 acres, or 486 fquare miles.

It is divided into *fix* baronies, *Half* RATHDOWN, NEWCASTLE, ARKLOW, BALLINACORR, TALBOTSTOWN, and SHILLELAGH; which contain † 58 parifhes, and 20 churches.—Of thefe, 49 parifhes and 17 churches are in the archbifhoprick of *Dublin,*—6 parifhes and 3 churches in the diocefe of *Leighlin*—and 3 parifhes with one church in that of *Ferns.*

The number of houfes in this county is 11,546, the inhabitants may therefore be computed at about 58,000; a very fcanty population for fo large an extent, as it amounts only to 23.75 in a fquare mile, and 26.9 acres to each houfe, on an average.

But a great part of WICKLOW is rendered unfit for habitation and incapable of culture, by mountains intermixed with rock and bogs. However, though the heart of the county be a cheerlefs wafte, the hills on the Eaft and Weft fides of it, and efpecially along the coaft, from 6 to 8 miles in breadth, being many of them well wooded and intermixed with profitable and fmiling vallies, form a delightful and various fcenery. They are crowded with gentlemen's feats, and are not without fmall towns and villages. The mountains of *Kippure,* near the county of Dublin, are the higheft, and very abrupt on the North fide.—*Keyden* on the border of Carlow, and *Sugar Loaf Hill* near Delgany, are alfo remarkable for height, and the latter for its conical form.

---

* From North to South 40½, from Eaft to Weft 33 Englifh miles, the area 500,600 Englifh acres, or 780 fquare miles.

† 5370 acres and 1000 fouls are about the average of each parifh.

I

In

In the mountainous part of this county many rivers have their
fources,—the *Liffey* with her tributary ftreams takes a circular
courfe through the county of Kildare, and falls into the bay of
Dublin. The *Slaney* runs Southward, and after croffing a part of
Carlow is received into the fea at Wexford. The *Fartrey* difembogues itfelf at *Wicklow*, and the *Ovoca* at *Arklow*.

There are no large towns in this county. WICKLOW, the
county town, is pleafantly fituated on a fmall harbour, and near a
beautiful ftrand abounding in fine pebbles, which is called the
Murrough.—The ale of Wicklow has been long celebrated in
Dublin. BRAY is reforted to by fea bathers in the fummer,
and like ARKLOW has a haven for fmall craft. On the Eaftern
fide of the county are BLESSINGTON, DUNLAVIN, BALTING-
LAS and CLONEGALL. In the neighbourhood of thefe Towns,
of RATHDRUM which is nearer the coaft, and of CARNEW in
the South, fome linen and much coarfe wollens are manufactured.

Not far from Rathdrum, at *Cronebane*, and alfo in the parifh of
Kileafhel, are extenfive *copper mines*. At Cronebane immenfe
quantities of copper are made, by fteeping bar-iron in the mine-
ral water, which entirely corrodes the original metal and fubftitutes the particles of copper in its place.

In the barony of SHILLELAH ftand the poor remains of a
foreft, once the moft celebrated in Ireland for the excellence
of its oak; which was exported to Britain and different parts
of Europe, and is ftill fhewn in the roof of Weftminfter-hall,
and of fome antient buildings on the continent, even at this
day.

There is fcarcely room to mention the antique ruins and
round towers of *Glandelough*, called the *Seven Churches*, which are
fituated in a deep valley, encompaffed with mountains; much lefs

to

to defcribe the many * natural beauties that attract the notice of the curious.

Ten members are returned to parliament for the *county* and *town* of *Wicklow*, the boroughs of *Bleffington*, *Baltinglas* and *Carysfort*, which laft is not even a village.

Obfervations on the Old Maps.

THE county is drawn too fhort from North to South by *two* miles. The courfes of rivers are in many places erroneous, the *Liffey* and *Bray-water* are made to unite, and fo are the rivers that fall into the fea at Wicklow and Kilcool. The boundary between the baronies of *Newcaftle* and *Ballinacor*, is improperly placed. The mountains are ill defcribed, and thofe of *Kippure* mifnamed *Stephenon*. The villages of *Glanteague*, *Togher* and *Ballinderry* are omitted.

WEXFORD.

THE length of the county of WEXFORD, which is fituated on the South of Wicklow, is from N. to S. † 44 miles, and the breadth from E. to W. 25. It contains 342,900 acres, or 535 fquare miles.

It is divided into eight baronies—GOREY, SCAREWALSH, BALLAGHEEN, BANTRY, SHELBURN, SHELMALIERE, BARGIE and FORTH, exclufive of the liberties of Wexford. Of 142 ‡

* The Glens of the Downs and Dunran, the Dargle, the Devil's Glen, the Waterfall at Powerfcourt, and many others.

† In Englifh meafure—length 56 miles, breadth 32, fuperficies 550,888 acres, or 695 fquare miles.

‡ Thefe parifhes would contain each at a medium 2400 acres and 750 fouls.

parifhes,

parifhes ; 140, containing 41 Churches, are in the bifhoprick of *Ferns*, and 2 parifhes with *one* church in the diocefe of *Dublin*.

The number of houfes in the county of Wexford is 21,040, and of inhabitants about 115,000 *, which is at the rate of 16.2 acres to a houfe, and 39.32 houfes in a fquare mile.

This county cannot be called hilly or mountainous, except on the frontiers of Carlow and Wicklow. Yet it contains a great deal of coarfe cold land, and ftiff clay foil, which the want of limeftone renders it difficult and expenfive to improve. But the baronies of *Bargie* and *Forth*, being of a lighter foil, are extremely well tilled, and produce large quantities of barley.

The river *Barrow* feparates Kilkenny from this county, and the *Slaney*, which croffes it from *Newtown Barry* to *Wexford*, affords a perpetual variety of picturefque and romantic views among its wooded and winding banks. *Lough Ta*, in the barony of Forth, receives into its bofom two or three fmart rivulets, but having no outlet, the waters accumulate and gradually overflow the adjacent grounds ; till the peafantry, once in three or four years, let them off, by making a cut through the high fand-bank that parts the lake from the fea, which very foon fills up again.

WEXFORD +, the fhire town, contains above 9000 fouls, and is fituated on a harbour which is large and beautiful, but too fhallow to admit veffels of great burden. Much coarfe woollen cloth is manufactured in this neighbourhood and about ENNISCORTHY, where there are alfo confiderable iron works, and fome trade ; the *Slaney* on which it is fituated being fo far navigable.—NEW ROSS, on the *Barrow*, which is there croffed by a ferry, is a pretty large town, and exports a great deal of beef and butter ; the river bringing up large fhips to

---

* At 5¼ per houfe, the average of Mr. Bufhe's return being 6.49.—See his paper *ut fupra*.

+ The number of houfes in this Town in 1788 was 1412.

the

the quay, with many articles for the confumption of the furrounding country. GOREY is but fmall. NEWTOWN-BARRY is one of the prettieft villages in the kingdom. From the very poor village of BALLYHACK there is a ferry to the fmall one of *Paffage* in the county of Waterford, acrofs the river *Suir*, which is there about a mile broad. The borough of CLONMINES is fallen into decay, fince the filver and lead mines in its neighbourhood have been neglected. Thofe of FETHERD and BANNOW are in the fame fituation. TAGHMON and FERNS are but very middling villages.

In the barony of FORTH the manners of the people differ in fome refpects from thofe of their neighbours—They have more induftry and cleanlinefs, and are much neater in their drefs—They fpeak no Irifh, but have among themfelves a language, which feems to be the Anglo-Saxon, but which falls very faft into difufe.

No fewer than *eighteen* members are returned to parliament for this *County*, the town of *Wexford*, and the boroughs of *Ennifcorthey, Gorey, New Rofs, Fetherd, Bannow, Clonmines*, and *Taghmon*.

Obfervations on the Old Maps.

They give the county *one* mile more in length than they ought. —The barony of *Bantry* is omitted in Jeffereys' map; nor is *Lough Ta*, in the barony of Forth, expreffed in any. The following villages are alfo omitted :—*Limbrick, Garrilough, Ram'sgrange, Broadway, Bridgetown, Killurin*, &c.

KILKENNY.

KILKENNY.

WEST of the county of Wexford lies that of KILKENNY, whose extent from N. to S. is 35 * miles, and from E. to W. 19. Its area measures 300,350 acres, or 469 square miles.

It contains nine baronies, exclusive of the *county of the city of* KILKENNY, and the *liberties* of the town of CALLEN; viz. FASSACHDINING, GALLMOY, CRANNAGH, GOWRAN, SHELLILOGHER, KELLS, KNOCTOPHER, IVERK, and the barony of IDA, IGRIN, AND IBERCON. These contain 127 † parishes, and 31 churches; of which 121 parishes, and 29 churches, are in the diocese of *Offory*; six parishes, with two churches, in that of *Leighlin*, and a very small part of one parish in the archbishopric of *Cashel*.

The number of houses is 17,569, which probably contain about 95 ‡ or 100,000 inhabitants. The houses are much fewer than what might be expected in a county which has been generally esteemed one of the most populous in Ireland; for there are, on an average, only 37.46 in a square mile, or one house to every 17 acres. This is below the medium of the provinces of *Leinster* or *Ulster*, but it somewhat exceeds the average of the whole kingdom; to which however it approximates nearer than any other county.

The soil is fruitful, and well tilled; and Kilkenny may be considered as one of the great corn counties.

---

* Length in English miles 44½, breadth 24; area 753 miles, and 482,464 acres.

† There are, on an average, 2364 houses, and about 700 souls, to every parish.

‡ The population was found to be 6.9 to a house, in a twentieth of the houses. *See Mr. Bushe's Tables in Mem. Roy. Ir. Acad.* 1789.

It

It is in general level, except in the barony of *Ida, Igrin, and Ibercon,* which is covered with rough hills from the neighbourhood of the *Suir,* till they terminate in *Brandon Hill,* a mountain in the barony of *Gowran.* No country can be better watered—The river *Barrow* forms its eaftern boundary— the *Suir* parts it from Waterford,—and both thefe rivers are navigable; as is alfo the *Nore* fo high as *Bennet's Bridge.* This river croffes the county from north to fouth, and receives in its courfe the *Dinin,* the river of *Callen,* and many other plentiful ftreams.

There are but a few towns of any note in this county, many boroughs, which were once confiderable, having dwindled into infignificant villages. CASTLE DURROW is a pretty little town, which lies in the midft of 2 or 3000 acres, that are infulated in the *Queen's County.* CASTLECOMER is but fmall, notwithftanding the vicinity of extenfive coal-mines, which not only fupply the furrounding country with fuel, but are conveyed in great quantities to very diftant parts of the kingdom, though by land carriage, being a hard ftone coal, particularly ufeful for fmiths' work. CALLEN is no longer of note, but its liberties, which extend near two miles round it, fhew that it was once confiderable. Of GOWRAN, KNOCKTOPHER, THOMASTOWN, URLINGFORD, FRESHFORD, NEWMARKET, &c. there is little to be faid. At BALLYSPELLIN, in *Galhnoy,* there is a chalibeate water, which would be more frequented, if there were better accommodation for ftrangers. The city of KILKENNY, and the borough of St. CANICE, or IRISHTOWN, form but one large town, which ftands in the midft of the *county of the city of Kilkenny.* It is delightfully fituated on the river *Nore,* over which are two handfome bridges.—Of the many large and good buildings that adorn this city, I fhall only mention the bifhop's
palace,

palace, the magnificent caftle of the Earl of Ormond, and the celebrated free-fchool or college founded by the Butler family, and lately rebuilt on a large fcale. The houfes are decorated with a very beautiful black and white marble, from the large quarries in the neighbourhood of this city, which fupply various parts of Ireland, and even London, with this commodity. KILKENNY contains about 16,000 * fouls, a large number for an inland country town; and has been long noted for the politenefs of its inhabitants.

In this city and its environs abundance of blankets and much coarfe woollen cloth are manufactured. In the hilly parts of the county there are great dairy farms, particularly in the neighbour-hood of *Waterford,* from whence large quantities of butter are exported.

Sixteen members are elected to the Houfe of Commons from this *county*—the city of *Kilkenny, St. Canice, Gowran, Knockto-pher, Thomaflown, Inniflioghe* and *Callen.*

Obfervations on the Old Maps.

The county meafures two miles longer and five miles broader in the old than in the new map—*Caftle Durrow* is placed in the Queen's County—the whole of the fouthern part of Kilkenny is erroneoufly reprefented as very mountainous—and the following villages are omitted: *Ballyfpellin, Johnflown, Pilltown, Ballyneale, Glanmore, Newmarket,* &c.

* The number of houfes in 1788 was 2689.

CARLOW.

CARLOW.

THIS fmall county, which is inferted as a wedge between the northern parts of Kilkenny and Wexford, meafures 26 * miles in length from N. to S. and 23, in the greateft breadth, from E. to W. It contains 137,000 acres, or 214 fquare miles, and is divided into *five* baronies and *fifty* † parifhes, 'which, with 13 churches, are all in the diocefe of Leighlin. The *Baronies* are, RAVILLY, CATHERLOGH, IDRONE, FORTH, and St. MULLINS.

The population of this county comes neareft to the average of the whole province; for it contains 8,763 houfes, and about 44,000 ‡ inhabitants, which will give, at a medium, 40.94 houfes to every fquare mile, and 15.6 acres to every houfe.

The river *Barrow*, which is navigable, runs through it from N. to S.—the *Slaney* croffes it alfo, in its courfe from Wicklow to Wexford.

That part of CARLOW which lies on the weft of the Barrow is covered with rough and high hills. Another mountainous tract continues all along the bounds of Wexford, beginning at the north, with the high and rocky *Mount Leinfter*, and terminating in that which is called *Blackftairs*, in the fouth. But the champaign country is extremely rich and fertile.

Of the Towns and Villages in this County, CARLOW, LEIGHLINBRIDGE, and TULLOW are the moft confiderable. CLONEGAL, HACKETSTOWN, PALATINETOWN, RUTLAND, GORE'S-BRIDGE, BURRES, &c. are fmall.

---

* The length is 23, the breadth 29 miles; and the area 220,098 acres, or 344 fquare miles, Englifh meafure.

† There are, on an average, 2,740 acres, and 880 perfons, to each parifh.

‡ At five per houfe—By the abftract it is 5.83. *See Mem. Royal Irifh Acad.* 1789.

K   The

The town of CARLOW is regularly built, and well fituated on the eaft fide of the *Barrow* (the weft end of the bridge being in the Queen's County), but has very little trade. LEIGHLIN BRIDGE is on the fame river, and feems to be in a more progreffive ftate of improvement.—The wooded hills which fkirt the river *Barrow* between thefe towns prefent a beautiful and varied fcenery.

Six members fit in parliament for this *County*, the town of *Carlow*, and the decayed borough of *Old Leighlin*.

Obfervations on the Old Maps.

I have drawn this county *one* mile longer from N. to S. and *one* mile narrower from E. to W. than the old maps make it— They omit *Mount Leinfter*, and the villages of *Palatine Town* and *Clonegal*.

KILDARE.

The county of KILDARE joins Carlow on the north, and ex- tends from N. to S. 32 * miles, and from E. to W. 21 ; form- ing an area of 236,750 acres, or 369 fquare miles.

The *ten* baronies, into which this county is divided, are— CARBURY, IKEATH AND OUGHTERANY, CLAINE, SALT, NAAS, GREAT CONNEL, OPHALY, KILCULLEN, NARRAGH AND REHBAN, KILKEA AND MOONE; in which are 113 † parifhes and 23 churches. Of thefe, 57 parifhes and 9 churches belong to the fee of *Dublin*, and 56 parifhes, with 14 churches, to that of *Kildare*.

* In Englifh meafure—Length 40½; breadth 26½; area 380,352 acres, or 593 fquare miles.
† The contents of every parifh would, on an average, amount to about 2,100 acres, and fomething lefs than 500 fouls.

The

The number of houses in this county is 11,205, and we may estimate the inhabitants at about 56,000 *. This gives, on an average, 30.36 houses to a square mile, and 21 acres to a house; which seems to be a very thin population for a county so near the capital, and not at all incumbered with mountains or high hills. But it must be observed, that large tracts of it are covered by the *Bog of Allen*, that upwards of 3000 acres are occupied by the *Curragh*; and that, although this beautiful plain affords pasture to an immense number of sheep, there are but a very few habitations scattered around its edges.

This county is full of springs and rivulets—The river *Barrow* forms its south-west boundary, and receives the *Grees*— The *Liffey* takes a circular course through the north-east of the county, and the river *Boyne* rises in the Bog of Allen. The *Barrow* is navigable from *Athy*, where it meets the GRAND CANAL, which, from Dublin, passes through this county, crossing the Liffey on an aqueduct bridge; and soon after branches off near *Claine*, in a collateral cut, to the Shannon. By this canal, not only merchandize and heavy goods are conveyed to and from the metropolis, but several boats, conveniently fitted up for the accommodation of travellers, pass daily between *Dublin* and *Monasterevan*.

In this county there are no large towns. Those which chiefly claim notice are, NAAS and ATHY †; at which the assizes are alternately held.

MONASTEREVAN, on the *Barrow*, is increasing in size and

---

* I have computed the number of inhabitants at the rate of only *five* to a house, as the abstract in Mr. Bushe's Paper states them at no more than 5.6. *Transf. Royal Irish Acad.* 1789.

† Having observed that English readers are commonly induced by the orthography to call this town *Athy*, I beg leave to remark, that in Ireland the name is always pronounced *Athy*.

trade, fince the completion of the canal has rendered it, in fome meafure, a centre of communication between Dublin and the more diftant parts of the kingdom.—KILDARE is chiefly fupported by the concourfe of nobility and gentry who attend the frequent races at the *Curragh;* which is the *Newmarket* of Ireland, and generally allowed to exceed the Englifh race-ground in elafticity of turf, and in characteriftic beauty.

At PROSPEROUS, a new village in the barony of *Claine*, great cotton works have been eftablifhed. BALLITORE is a very pretty village.

Ten members reprefent in the Houfe of Commons this *County*, the towns of *Kildare*, *Naas* and *Athy*, and the borough of *Harriftown*, which confifts only of a fingle houfe.

Obfervations on the Old Maps.

They exceed in the breadth of the county *two* miles, and are not quite correct in the boundary and divifions. *Jeffereys* mifcalls Monaftereven *Monflere*. The fituations of *Cappoge* and of *Profperous* are not marked.

* * *

QUEEN'S COUNTY.

The QUEEN'S COUNTY is fituated on the fouth-weft of Kildare, from which it is partly divided by the river *Barrow*, and is of a very compact form ; being 25 * miles in length, and as many in breadth. It contains 235,300 acres, or 367 fquare miles, and is divided into eight baronies—PORTNEHINCH, TINEHINCH,

* Nearly 32 Englifh miles.—The area, in the fame meafure, is 378,023 acres, or about 590 fquare miles.

UPPER

UPPER-OSSORY, MARYBOROUGH, STRADBALLY, BALLY-
ADAMS, CULLINAGH and SLEWMARGY.

Of the 50* parifhes, and 26 churches which they comprehend,
27 parifhes and 14 churches are in the bifhopric of *Leighlin* ; 14
parifhes and 6 churches in that of *Offory*; 7 parifhes, with 6
churches, in the diocefe of *Kildare*; *one* parifh in that of *Killaloe*, and part of one in *Dublin*.

This county is more populous than any of the preceding five;
for it contains 15,048 houfes, which on an average would give
41 houfes to a fquare mile, and 15.6 acres to a houfe. The
number of inhabitants will amount to upwards of 82,000, at $5\frac{1}{2}$
to a houfe.

There are in this county extenfive tracts of bog, and a good
deal of cold wet ground, efpecially near the mountains ; yet the
greater part of it is well cultivated, and fome places rich and
beautiful. It is a very level country, except in the fouth of the
barony of *Slewmargy*, and on the borders of the *King's County*.

The high and fteep mountains of *Sliebh-bloom* † form fo im-
practicable a barrier between the two counties, that in a range of
fourteen miles, they afford but one, and that a very difficult and
narrow pafs into the King's County, called the *Gap of Glandine*.
In this great ridge are the fources of the *Barrow* and the *Nore* ‡ ;
the *Barrow* running North-eaft to Monafterevan, where it
changes its direction to the South, and the Nore croffing the
Queen's County by a fouthern courfe into Kilkenny.

MARYBOROUGH, the county town, is not large: but in its

---

* Every parifh contains, on an average, 4,700 acres, and 1,640 perfons.

† Thefe mountains are alfo named *Ard-na-Erin*, which, in the Irifh language, fignifies in
the Height of *Ireland*.

‡ As thefe two rivers unite with the *Suir*, near Waterford, and as that river rifes in the
mountain of *Bendbú* in Tipperary, which is at no great diftance from the fouthern extremity of Sliebh-bloom, they are not improperly called the *Three Sifters*.

neigh-

neighbourhood is manufactured a great quantity of fluffs, ferges, druggets, and other woollen goods. The fame manufactures are carried on at MOUNTMELLICK and MOUNTRATH; in which towns the wool-combing bufinefs flourifhed exceedingly, till within a few years, when the demand for worfted yarn from Norwich and other parts of England ceafed. Forges and furnaces for iron have long been eftablifhed about *Mountrath*; but charcoal is become fo fcarce that, of late years, they have not been often at work. BALLYNAKILL is a fmall town, and STRADBALLY a very pretty clean village; but PORTARLINGTON (of which a fmall part is in the King's County), is a confiderable town, full of gentry, and noted for many large fchools, principally employed in the preparative education of very young children.

There is an extenfive colliery at *Dunane* in *Slewmargy*, whofe coals are preferred to thofe of Caftlecomer, and all the hills of that part of the county are full of this kind of coal. Much cheefe is made in this and the next county, which brings a good price in Dublin.

The Queen's County deputes *eight* members to parliament, who reprefent the *County, Maryborough, Portarlington*, and *Ballynakill*.

Obfervations on the Old Maps.

They make this county *three* miles too broad from E. to W. and reprefent *Slewmargy* as a flat country. Neither the mountains of *Sliebh-bloom*, nor the courfe of the *Barrow*, are correctly delineated. The villages of *Cafletown* and *Dunane*, both in Slewmargy, are omitted.

KING'S

KING'S COUNTY.

THIS County bounds the Queen's County on the north and weſt. Its length from N. to S. is 34 miles; and the breadth from E. to W. which, in the northern and broadeſt part, is 32, diminiſhes gradually to a very narrow compaſs as it ſtretches to the Southward. The KING'S COUNTY contains 282,200 acres, which make upwards of 440 ſquare miles *; and is divided into eleven baronies, viz.—WARRENSTOWN, COOLESTOWN, PHILIPSTOWN, BALLYCOWEN, KILCOURSEY, GARRYCASTLE, GESHIL, BALLIBOY, EGLISH or FIRCAL, BALLYBRITT and CLONLISK.

In theſe there are 16 pariſhes and 10 churches, under the ſee of *Meath*; 16 pariſhes, with 6 churches, under that of *Killaloe*; 18 pariſhes and 8 churches in the dioceſe of *Kildare*; *one* pariſh in Clonfert, and *one* inſulated pariſh, with its church, dependant upon the ſee of *Oſſory*; in all 52 † pariſhes and 25 churches.

The number of houſes in this county amounts to 13,536, which will give 30.76 houſes to a ſquare mile, and 20.8 acres to a houſe on an average; and may contain about 74,500 inhabitants ‡.

If this ſhould be eſteemed a thin population, it muſt be obſerved, that the Bog of Allen covers a great portion of this county, and that ſome part of it is rendered uninhabitable by the moun-

* In Engliſh meaſure, the length is 43 miles, and the breadth 39; the area 453,370 acres, or 707 ſquare miles.

† There are on an average 5,400 acres, and 1,430 ſouls to a pariſh, through this county.

‡ At 5½ per houſe; the average population appearing, by the return in Mr. Buſhe's Paper, to be 6.11 to a houſe. *Mem. Royal Iriſh Academy*, 1789.

tains

tains of *Sliebh-bloom*. This range of mountains excepted, the King's County may be confidered as a level country. In many places the foil is deep and rich, and in fome parts well cultivated; but much of the barony of *Garrycaftle* ftill remains in a very naked and unimproved ftate. In this wild country, on the banks of the Shannon, ftand the ruins of *Clonmacnois*; ufually called the *Seven Churches*, from the many veftiges of religious buildings, among which two round towers only remain free from decay.

The *Shannon* forms the weftern boundary of the County for many ·miles, and the little *Brofna*, which falls into that great river, divides it from Tipperary; while the larger *Brofna*, after winding through a great part of it, between ·pleafant banks, lofes itfelf alfo in the *Shannon*. The *Boyne* and the leffer *Barrow* water the fkirts of it. There are, befides, feveral fmall rivers and fome lakes, of which *Lough Pallis* and *Lough Annagh* are the largeft; and the *Grand Canal* croffes the Northern part of the county.

Of the towns in the King's County, BIRR is the moft confiderable. It was formerly called *Parfonftown*, and has very lately refumed that name, by act of parliament. At BANAGHER there is an excellent endowment for a fchool, and a very ancient bridge over the *Shannon*. Six miles higher up, at the village of RAGHERA, a very noble one has been erected within a few years. TULLA-MORE, on the river *Clodagh*, is a pretty little town, in which, as well as in BIRR, there are many gentlemens houfes, and fome manufactures, which the advantages of the Grand Canal will probably improve. In the neighbourhood of CLARA, which is fituated on the *Brofna*, much linen is made. At PHILIPSTOWN the affizes are held, but it is a very indifferent place. FRANK-FORD and BALLYBOY on the *Silver* river, FERBANE upon the *Brofna*, BALLYCOWEN and CLOGHAN, KILLEIGH and GESHIL are fmall; but EDENDERRY is a thriving town, chiefly inhabited

bited by induſtrious quakers, and contiguous to the weſtern cut of the Grand Canal.

Six members repreſent, in the Houſe of Commons, this *County* and the towns of *Philipſtown* and *Banagher*.

Obſervations on the Old Maps.

They exceed the juſt dimenſions of the county from E. to W. three miles, and they omit the villages of *Durrow*, *Clara*, *Ballycumber*, *Ballycowen*, *Frankford*, and ſome others.

WESTMEATH.

North of the King's County lies WESTMEATH, whoſe greateſt extent from E. to W. is 33 miles, and from N. to S. 27, and whoſe area meaſures 231,538 acres, or 361 ſquare miles *.

There are *twelve* baronies in this county—CLONLONAN, MOYCASHEL, FARTULLAGH, FARBILL, MOYASHEL AND MAGHERADERNON, DELVIN, *Half* FOWRE, MOYGOISH, CORKERRY, RATHCONRATH, KILKENNY-WEST, and the *territory*, as it is called, of BRAWNY.

They comprehend 62 † pariſhes and 21 churches; 59 pariſhes and 20 churches in the dioceſe of *Meath*, and three pariſhes, with *one* church, in the biſhopric of *Ardagh*.

The population may be computed at 69,000 perſons, dwelling in 13,693 houſes: and if they were equally diſtributed throughout the county, every ſquare mile would contain 37.93 houſes; to each of which there would be 17 acres.

---

* From E. to W. 42, from N. to S. 34 miles; 371,979 acres, or 577 ſquare miles, Engliſh meaſure.

† At a medium, each pariſh would contain about 3700 acres, and 1100 perſons.

L No

No part of this county is embarraffed with mountains, but a great number of acres are rendered unproductive by large lakes and extenfive bogs; yet the convenience of fuel, the abundance of gravelly hills, and the variety of profpects which arife from thefe beautiful lakes, and the undulating form of the furface, render it a very pleafant and healthful country. The foil is in general light, but in fome places deep and rich; and though there is more of it kept under grafs than employed in tillage, yet the plough is by no means neglected; for, after fupplying the home confumption, the farmers of this county largely contribute to the exportation of oats from Drogheda.

No county can be better watered. The *Shannon* divides it from Connaught: the river *Inny* forms the greater part of its north-weft boundary, paffing through feveral lakes: the *Dele* croffes the eaftern fide of it, and the *Brofna*, which iffues from *Lough Hoyle*, runs fouthward into the King's County. It is fingular, that from this lake two rivers flow, in oppofite directions: the *Brofna* takes a fouthern courfe; while a fhort and rapid ftream runs weftward into *Lough Iron*, which difcharges its waters by the *Inny* into the *Shannon*. Lough *Lene*, Lough *Iron*, Lough *Derveragh*, Lough *Hoyle*, and Lough *Ennel*, are the principal lakes. To thefe muft be added that large and beautiful expanfion of the *Shannon*, which is full of wooded iflands, and called Lough *Ree*, or the Royal Lake.

The towns in Weftmeath are neither large nor numerous.— ATHLONE is the moft confiderable; fituated on the Shannon, it was formerly an important pafs into the Weftern Province, and is the moft central town in the ifland. MULLINGAR, the county town, is well fituated between the two lakes *Hoyle* and *Ennel*, and is noted for a very confiderable horfe-fair, and a great wool-fair. MOATE-GRENOGUE, KILBEGGAN, KINNEGAD and CAS-TLEPOLLARD, are fmall. In this county alfo, the linen manu-facture makes a confiderable progrefs.

From

From *Weſtmeath*, *Mullingar*, *Athlone*, *Kilbeggan*, and the now miſerable village of *Foxre*, *ten* members are deputed to parliament.

Obſervations on the Old Maps.

The county is repreſented in them three miles too broad—Several baronies are miſnamed, and *Mullingar* is erroneouſly made a thirteenth barony. The villages of *Collinſtown*, *Clonmellon*, *Beggar's Bridge*, *Miltown*, and *Horſeleap*, with the hill of *Uſneah*, are omitted.

LONGFORD.

Adjoining to Weſtmeath on the north, the county of Longford extends in length from N. to S. 20 miles ; in breadth, from E. to W. 19 ; and contains 134,152 acres, or 209 ſquare miles *.

This ſmall county is divided into ſix baronies and 23 pariſhes †, which compriſe 16 churches. The baronies are, LONGFORD, GRANARD, ARDAGH, MOYDOH, RATHLINE, and SHROWLE. Of the pariſhes, 22, containing 15 churches, are in the biſhopric of *Ardagh*, and *one* pariſh, with its church, in the diocefe of *Meath*.

LONGFORD is extremely well peopled ; for, notwithſtanding the bleak and rough hills in the northern angle, and the large bogs that are ſpread over the ſouth-weſt and other parts of the county, it comprehends 10,026 houſes ; of which, if they were equally diſtributed, there would be 47.97 in every ſquare mile,

---

* Length 25 miles, breadth 24 miles ; contents 215,522 acres, or 336 ſquare miles, Engliſh meaſure.

† The pariſhes contain, on a medium, 5800 acres, and about 2200 ſouls.

with

with only 13.4 acres to each houfe. At five fouls to a houfe, the County muſt contain upwards of 50,100 inhabitants *.

By far the greateſt part of it is flat, and in ſome places ſubjeƈt to be overflowed; yet the farmers are able to ſend large cargoes of oats to the port of Drogheda. Many hands are employed in ſpinning and weaving; much linen is made in this county, and great quantities of yarn are ſent to more diſtant markets.

The *Shannon* forms the weſtern boundary of this County. The *Inny* flows at the ſouth, Lough *Gawnagh* expands its waters over many miles in the north; and the *Camlin* and *Fallen*, with other ſmaller rivers, ſupply the heart of it.

The town of LONGFORD, on the *Camlin*, is of middling ſize, and pretty well built. GRANARD, EDGEWORTHSTOWN, BALLYMAHON, LANESBOROUGH, ST. JOHNSTOWN, &c. are ſmall.

Ten members however are deputed to parliament by this *County*, and the boroughs of *Longford*, *Lanesborough*, *Granard*, and *St. Johnſtown*.

Obſervations on the Old Maps.

The length of the County is too great in the old maps by *three* miles, and the breadth by *two*. The courſe of the *Inny* through the barony of Shrowle is quite erroneous, and the names of ſeveral baronies are ill ſpelled. The villages of *Kenagh*, *Barry*, and *Firmount*, are not in Jefferys' Map.

---

* In the report made to the Revenue Board, the proportional population is at the rate of 5.87 to a houfe. *See Mem. Royal Iriſh Academy.*

PROVINCE of CONNAUGHT.

THIS Province comprehends *five* counties, which are fubdivided into 42 baronies, and 296 parifhes. The number of acres
in Connaught being 2,630,300, which make about 4108 fquare
miles, and the number of houfes only 95,821; there are 27.4
acres to every houfe, and only 23.31 houfes in a fquare mile, on
an average of the whole province.

ROSCOMMON.

This County, which the *Shannon* divides from Longford and
Weftmeath, extends in length from N. to S. 47 miles; the
breadth varies confiderably, and in the broadeft part, about the
middle of the county, is 29 miles. The area meafures 346,650
acres, which are equal to 541 fquare miles *.

The fix baronies of BOYLE, BALLINTOBAR, *Half* BALLI-
MOE, ROSCOMMON, ATHLONE, and MOYCARNE, contain 56 †
parifhes and 22 churches; 50 parifhes, with 20 churches, under
the fee of *Elphin*; three parifhes and one church under that of
Tuam; two parifhes, with a church, in the diocefe of *Clonfert*;
and *one* parifh in the bifhopric of *Ardagh*.

Of all the counties weft of the Shannon this is the beft peopled; yet, as it contains only 17,137 houfes, there is on a medium

---

* In Englifh meafure, the length is 60 miles, the greateft breadth 37, and the area
556,847 acres, or 869 fquare miles.

† Thefe parifhes contain each, on an average, about 6,200 acres, and near 1,600
fouls.

5   but

but one houfe for every 20.1 acres, and only 31.86 to a fquare mile. The inhabitants may be eftimated at about *eighty-fix thoufand.*

Rofcommon is a flat open country, in fome places fprinkled with rocks, in many interrupted by extenfive bogs, and but little diverfified with hills. The only mountains within the county are in the parifh of *Kilronan*; a nook between *Lough Arrow* and *Lough Allen*, and thefe are become valuable on account of the coal and iron which they are found to contain. The lofty *Curlew Mountains*, which join Lough Arrow, feparate this county from Sligo. In the plains of Rofcommon the foil is rich, and as fit for the ploughman as the grazier. There is however but a fmall part under tillage, in comparifon of what is devoted to the breeding of black cattle and fheep.

The river *Shannon* winds along the eaftern boundary, branching, in a courfe of fifty miles, into feveral fine lakes; of which Lough *Ree*, Lough *Baffin*, and Lough *Allen*, are the largeft. The river *Suck* divides this county from *Galway* for a great length of way, till it lofes its name and waters in the Shannon; and many other ftreams and fmall lakes fertilize and enliven the fields. The largeft of the lakes is Lough *Key*, in the North of the County, which is rendered delightful by wooded iflands and furrounding groves. There are no towns of great fize or confequence in Rofcommon—The principal are BOYLE, where the linen market is of late become very confiderable, from the increafe of yarn and linen manufactures in the neighbourhood— ROSCOMMON, the County town—STROKESTOWN, ELPHIN, CASTLEREAGH, MILTOWN-PASS, &c. Part of *Athlone* is alfo in *Rofcommon*.

This *County*, the boroughs of *Rofcommon*, *Boyle*, and *Tulfk*, which laft is a wretched village, are reprefented in parliament by *eight* members.

                                                            Obferva-

Obſervations on the Old Maps.

They reprefent the County one mile ſhorter and two miles broader than the new map. The *river* which flows from Lough *Gara* into Lough *Key* is not expreſſed. The towns of *Ballimee* and *Ballinaſtoe* are placed in *this* County, though they are both in *Galway*; and the villages of *Drumda*, *Ardcarn*, *Ballyfernon*, *Ballintra*, and *Sandfield*, &c. are omitted.

LEITRIM.

North-weſt of Rofcommon, with the Shannon intervening, lies the county of LEITRIM, which extends from the county of Longford to Donegal Bay, 41 miles in length. In form ſomewhat like an hour-glafs, it varies greatly in breadth, being in the wideſt parts 16 and in the narroweſt only ſix miles acrofs. It contains 255,950 acres, or about 400 fquare miles *.

This county is divided into *five* baronies and 17 † parifhes. The baronies are, MOHILL, LEITRIM, CARIGALLEN, DROMAHAIRE, and ROSSCLOGHER. *Ten* parifhes are in the diocefe of *Kilmore*, and 7 in *Ardagh*. There are 8 churches in each.

The number of houfes in LEITRIM being 10,026 (which probably contain upwards of 50,000 inhabitants), if they were equally diftributed throughout the county, there would be in

* The length of Leitrim, in Englifh meafure, is 52 miles; the greateſt breadth 20, and the leaſt 7½. The area contains 407,260 acres, or 652 fquare miles.

† Every parifh comprehends, on an average, upwards of 15,000 acres, and fomewhat lefs than 3000 fouls.

very

every fquare mile 27.48 houfes, with 23.2 acres to each. But the two northern baronies are by no means fo populous as the other three. The mountains of *Sliebh-Anezer* and *Dartry* cover almoft the whole of *Roffelogher ;* and a large fcope of *Dromahaire* is occupied by *Sliebh-an-Erin* and other mountainous groups. But thefe great hills are far from unprofitable ; for, producing abundance of coarfe grafs, they annually pour forth immenfe droves of young cattle. The fouthern baronies are level, and their foil good : agriculture of courfe improves, and population increafes with the linen bufinefs, which has made great progrefs in a country fo well adapted to the growth of flax, and fo con‑ venient to the manufacturer in point of fuel and water.

It abounds in fmall rivers, and lakes—The largeft of thefe is *Lough Allen*, a deep water, 8 miles in length, and 2½ in breadth. On the weftern banks of this lake the hills teem with coal and iron, in this county, as well as in Rofcommon. Great iron works have been lately eftablifhed at *Arigna*, and as foon as the com‑ pletion of the Royal Canal opens a communication between Dublin and the Shannon, the valuable products of this diftant county will find an eafy conveyance to market. About four miles north of *Lough Allen*, the *Shannon* iffues from *Lough Clean*, a fmall lake, which is confidered as the fountain of that noble river. This lake is not four miles diftant from the river *Bonnet,* which carries boats into Lough Gilly, and from thence into Sligo Bay. Perhaps the day may come, when the fpirit of enterprize and commerce will open itfelf a paffage by this channel alfo.

The towns and villages in Leitrim are very fmall—Mohill, Manor-Hamilton, and Dromahaire, are perhaps the beft. —Carrick is the fhire town. Near Drumsnaw, a neat vil‑ lage, charmingly fituated on the wooded banks of the Shannon, there is a *chalibeate* fpring, whofe medicinal virtues have been found very great.

Six

Six members reprefent in parliament the county of *Leitrim,* and the boroughs of *Carrick* and *Jameſlown.*

Obfervations on the Old Maps.

In the length of the county they fall ſhort *a mile,* and in the breadth they exceed as much.—The *villages* of *Kinlough* and *Drumkeirn* are omitted. *Drumſhambo, Ballintra,* and *Keſhcarrigan* are mifplaced by Bowles; and the two laſt are omitted by Jefferys.

SLIGO.

THE county of SLIGO lies on the weſt of Leitrim, and on the north of Rofcommon. Its greateſt length, from *Bunduff* in the north, to the *Curlew Mountains* in the ſouth, is 31 miles; and the greateſt breadth 29. It contains 247,150 acres, or 386 ſquare miles*, and is divided into *ſix* baronies, CARBURY, TYRERAGH, LENEY, CORRAN, COOLAVIN and TIRAGHRILL; which comprife † 39 pariſhes and 16 churches. Of thefe, 16 pariſhes and 3 churches are in the biſhoprick of *Elphin;* 14 pariſhes and 6 churches in that of *Achonry,* and 8 pariſhes with 6 churches, in the diocefe of *Killalla.* *One* pariſh and its church belong to the fee of *Ardagh.*

In this county there are 11,509 houfes, the average of which is 29.81 to a ſquare mile, or one houfe to 21.5 acres. And the number of inhabitants may be about 60,000.

* The dimenſions of this county are, in Englifh meafure,—length 39½, and breadth 37 miles. Its contents 397,060 acres, or 620 fquare miles.

† The 39 pariſhes, contain on an average 6,360 acres, and 1,280 fouls each.

M The

The county of *Sligo* contains very good land, intermixed with large tracts of coarse and unprofitable ground. In the barony of *Carbury*, are the mountains of *Benbulb* and *Samore*. A chain of rough hills extends from Lough *Gilly* to the bounds of Roscommon and Leitrim. *Tyreragh*, though level along the coast, is interfected by large bogs; and the southern part of it is bounded by the *Ox Mountain*, *Sliebh Dham*, and a great range of defolate hills, that extend a good way into the barony of *Leny*, in which also there is a great scope of bog. The *Curlews*, and other mountains, cover the most of *Coolavin*; and *Kifhcorran* forms a long ridge on the borders of *Tyraghrill*. *

Among these hills there are many large lakes and abundance of rivers.—The *Moy* rises in the mountain of *Knocknafhee*, and after receiving the waters of Lough *Calt* and Lough *Conn*, flows in a broad stream to the bay of Killalla. Lough *Arrow* is about eight miles long, full of islands, and of a very irregular form. A river of the same name proceeds from it, and running northward, to *Ballyfadere*, rushes at once into the fea in a stupendous cataract. *Lough Garra* is also an extensive lake.

Lough Gilly exhibits that variety of charming prospects which bold hills, wooded lawns, and large islands cloathed with verdure and crowned with trees, united with a great extent of water, cannot fail to produce. Upon the river by which the waters of this lake are discharged into a large bay, stands the town of SLIGO, and vessels of two hundred tons come up to the quays. The trade of this town has been increasing for some years, and the number of inhabitants are estimated at upwards of 8,000 †. There is no other town of note in the county, but there are many small villages; in which, and the surrounding

* On the summits of most of these mountains there are very large *Cairns* or *Carnedhs*.

† The number of houses in Sligo, at the end of 1788, was 916. *Tranfact. of the Royal Irish Academy, for* 1789.

country, the linen bufinefs wears a flourifhing afpect, efpecially in the vicinity of *Ballymote*.

This *county* and *town* of SLIGO are reprefented in parliament by *four* members.

Obfervations on the Old Maps.

This new map deviates confiderably from the old in the fize of the county, which, by the prefent conftruction, is made 5 miles fhorter from eaft to weft, and *one* mile lefs from north to fouth.—The line of coaft is alfo varied, and the interefting ifland of *Inifmurry*, placed in its true pofition, according to M'Kenzy's charts.—In the old maps, the boundary between the baronies of *Corran* and *Tiraghrill* is incorrect ; and that part of *Coolavin*, which is on the eaft of Lough *Garra*, is given to the county of Rofcommon. The mountains are very ill defcribed throughout the whole county, and Lough *Calt* is omitted. Neither do they mark the fmall villages of *Efky*, *Skreen*, *Tobarcorry*, *Tobarfcanavan*, *Ballinode*, *Courtftrand*, *Acharrow*, and *Liffadill*, which laft is celebrated for the excellence of its oyfters.

MAYO.

THIS county, which joins Sligo, is bounded on the north and weft by the ocean. Extending from N. to S. 49 miles, and from E. to W. 45, it is exceeded in dimenfions by Cork and Galway only ; for it contains 790,600 acres, or 1,235 fquare miles. *

Mayo is divided into nine baronies—TYRAWLY, GALLEN, COSTELLO, CLANMORRIS, KILMAIN, MORISK, CARRAGH,

---

* The dimenfions of Mayo, in Englifh meafure, are—length 62 miles ; breadth 57 ; area 4,270.144 acres, or 1,984 fquare miles.

BUR-

BURRISHOOLE, and ERRIS. Of 68 * parishes, and 20 churches,
which they comprise, 37 parishes and 12 churches are in the
diocese of *Tuam*,—17 parishes with 6 churches in *Killalla*,—
13 parishes and 2 churches in *Achonry*, and *one* parish in the
bishoprick of *Elphin*.

The number of houses in this county is 27,970. This is at the
rate of 22.64 in a square mile, and of 28.2 acres to a house,
which is near the medium of the province.—The number of
inhabitants may be estimated at 140,000. †

The soil of the county of Mayo varies prodigiously—from the
bleak and rugged mountain to the fertile and chearful plain.
The baronies of *Kilmain* and *Clonmorris*, the greatest part of *Car-*
ragh, *Costello*, and *Gallen*, and a large portion of *Tyrawly*, are
arable and champaign; and though not yet arrived at a high degree of cultivation, they produce a sufficiency of corn and flax, for
home consumption, and supply other counties with abundance of
fat and store cattle. In the mountainous barony of *Burrishoole*
there are some fruitful grounds, along the coast and in the vallies.
But the large barony of *Erris*, and the western part of *Tyrawly*,
are overspread with an immense mass of uninhabited mountains,
and trackless bogs, without roads and very difficult of access, to
the few farmers and fishermen who dwell upon the coast, and to
the inhabitants of the *Mullet* ; a peninsula, which is said to be
fertile, pleasant, and well inhabited ‡.

Among the mountains that cover *Morisk*, *Croagh-patrick* (or
Crowpatrick) claims the pre-eminence, the conick summit of
which is distinguished at a vast distance, rising 2,666 ‖ feet

---

* The 68 parishes comprehend each on an average about 11,600 acres and 2,000 souls.

† In the returns of Mr. Bushe, the number of inhabitants is stated at 5.8 per house.
See Trans. Roy. Irish Acad. ut supra.

‡ When I was in Mayo, the season was unfortunately too much advanced, for to ven-
ture so far into this difficult country.

‖ The height of *Mangerton*, in Kerry, is about 2,500 feet.

above

above the level of the fea, and being generally efteemed the higheft mountain in Ireland. * Mount *Nephin*, though little inferior to it in height and fublimity, being 2,640 feet high, is of a very different character : for it ftands almoft infulated, and appears rounded on all fides and at top, like a huge rath or barrow.

There are, in the flat country, that borders upon the lakes of *Maſk* and *Carrah*, many miles of rocky ground, which, at a diftance, appear like one immenfe fheet of white ftone. But upon a nearer infpection of thefe fingular rocks, they are perceived to ftand in parrallel lines, from one to three feet, above the furface, like flag-ftones pitched in the ground upon their edges; and, however they may vary in fhape, fize, and diftance, they are all calcareous, and have all the fame direction. Fiffures of a great depth are found in fome of the narroweft interftices : but in general, the verdure between them is beautiful, and the pafture excellent for fheep. Large caverns and fubterraneous waters are alfo frequent in this part of the country, efpecially near CONG. At the back of that fmall village, a very broad river rufhes at once from beneath a gently floping bank, and after a rapid courfe of about a mile, lofes itfelf in *Lough Corrib*. It is fuppofed to be the outlet of a fub-terraneous channel, through which the fuperfluous waters of Lough *Maſk* and Lough *Carrah* are difcharged into *Corrib*. This rocky part of Mayo abounds alfo with *Turlachs*, as they are cal-led in Irifh. Thefe are plains, fome of them very extenfive, which having no vifible communication with any brooks or rivers, in the winter are covered with water, and become in the fummer a rich and firm pafturage, the waters rifing and retiring through rocky clefts in the bottoms.

There are many fine lakes in this county. Lough *Conn* at the foot of Mount Nephin is nine miles long : Lough *Maſk* is longer by two miles, and confiderably broader.

* At the top of this pinnacle, is a very large and remarkable *Cairn*.

There

There are many noble harbours on the coaft of Mayo : *Kil-lalla Bay*, at the mouth of the river Moy, *Broadhaven* and *Black-ſod* Bay, between the *Mullet* and the main, are much larger, but ſtill leſs frequented. *Clewbay*, ſheltered on the north and ſouth by the mountains of Burriſhoole and Moriſk, and defended from the weſtern ſtorms by the high and rocky Iſle of *Clare*, affords a deep and ſafe anchorage among the innumerable iſlands that adorn the bottom of this magnificent haven. At the *Killeries*, a large bay and an excellent harbour ſhelter, in the fiſhing feaſon, a vaſt number of herring buſſes, which rendezvous there, from all parts of Galway and Mayo.

To this county belong the great iſland of *Achil*, and the ſmall ones of *Achil-beg*, *Anagh*, *Inis-turk*, *Cahir*, *Iniſtegil*, &c.

CASTLEBAR, the ſhire town, and the moſt conſiderable in Mayo, has been very much enlarged within a few years, and is ſtill increaſing in ſize and opulence, by the judicious encourage-ment which Lord Lucan gives to the linen manufacture, and to other trades. BALLINROBE, which is much ſmaller, is alſo in a proſperous ſtate. NEWPORT-PRATT, is a ſmall ſea-port near the mouth of a fine river, at the north-eaſt of *Clewbay*. WEST-PORT on a beautiful bay, wooded to the water's edge, in the ſouth-eaſt angle of the ſame great haven, is a ſmall new town, neatly built, and daily improving, under the auſpices of Lord Al-tamont. KILLALLA, though a biſhop's ſee, and ſituated on a fine harbour, is but a poor town. FOXFORD is alſo a wretched place. BALLINA being connected with *Ardnaree*, in the county of Sligo, by a bridge over the *Moy*, they form but one town, which is neat and thriving, and has a briſk market for linen every week —There are in this county ſeveral other ſmall towns or villages, ſuch as HOLLYMOUNT, KILMAIN, MAYO, BALLY-HAUNIS, MANILLA, BALLAGH, BALLCARRA, &c.

In the lakes of this county there are, beſides abundance of large trout, ſalmon, and other kinds of fiſh, a ſpecies of trout, called

the

the *Gilleroe*, whofe ftomach has the appearance and confiftency of a gizzard: the fifh is excellent, and this *gizzard* is efteemed a great delicacy. At *Turlogh*, in the barony of *Carragh*, there is a quarry of beautiful marble, as black as jet, and free from any mixture of white or grey, but as yet very little worked.

Large as this county is, *four* members only are deputed to parliament by the *County* and the town of *Caftlebar*.

Obfervations on the Old Maps.

The whole coaft is incorrectly drawn in the old maps, particularly the *Killery* harbour; of which the form is very erroneous, and the name omitted. *Clewbay* is not accurate, and *Clare* Ifland is mifplaced: fo is *Achilbeg*, which they name *Kildanat*. The peninfula of *Coraan* is feparated from the main land and made an ifland, and with *Achil*, which is very ill-fhaped, is called *Achil Iflands*. The ifles of *Inifkea* are omitted—*Black-Sod Bay* is called Black Harbour, and *Black-Sod Point* Saddle Head—Lough *Conn* is ill-fhaped, and Lough *Cullin* omitted. The courfe of the *river* at Ballinrobe is wrong—The mountain of *Nephin* is ill reprefented, and *Crowpatrick* is too far from the fhore. Jefferys places the town of *Newport* at the fouth, inftead of the north angle of *Clewbay*, and fubftitutes for it the imaginary one of *Broca*—The barony of *Morifk* is erroneoufly called *Joyces Country*, which is in the county of Galway—The fmall village of *Dunkeehan* they call *Sargala*, and omit thofe of *Portachloe*, *Inver*, *Lettikeen*, *Clare*, *Manilla*, &c.

GALWAY.

This great county, the fecond in fize, but the laft in population of the thirty-two, lies immediately fouth of Mayo, and extends

tends 43 miles from N. to S. and 76 from E. to W.; containing 989,950 acres, which make 1,546 fquare miles *.

It is divided into *fixteen* baronies, exclufive of the *Liberties of* GALWAY, and contains 116 † parifhes, and 28 churches.

Of the parifhes, 49, in which are 11 churches, are under the archbifhop of *Tuam*; 37, with 9 churches, in the bifhopric of *Clonfert*; and 20, with 4 churches, in that of *Kilmacduagh*; 8 parifhes, and 3 churches, are in the diocefe of *Ephin*; and two parifhes, with a church, in *Killaloe*. The 17 baronies are, CLARE, DOWNAMORE, *Half* BALLIMOE, KILLIHAN, TIA-QUIN, ATHENRY, KILLCONNEL, CLONMACOW, LONGFORD, LETRIM, LOUGHREA, KILTARTAN, and DUNKELLIN.

The number of houfes in the county of GALWAY is 28,212, which may contain about 142,000 fouls ‡. On an average of the whole county, there are 35 acres to a houfe, and but 18.24 houfes to a fquare mile. This very fcanty population may be in fome meafure accounted for, by the rude flate of the three baronies on the weft of *Lough Corrib*, which amount to a third part of the county; as they contain about 341,600 acres: the lake itfelf covers 31,300. The extenfive country on this fide of the lake is flat; with the exception of a few fertile hills of no great height, and fome low mountains on the borders of *Clare*. The foil is warm and fertile, covering, at no great depth, a ftratum of lime-ftone rock, which in the baronies of *Dunkillen* and *Kiltartan*, and in many other places, rifes fo thick above the furface as to render thofe parts unfit for tillage, though they are excellent

---

* In Englifh meafure—Galway extends from N. to S. 54¼ miles; from E. to W. 96¼; and contains 1,739,591 acres, or 2,718 fquare miles.

† The parifhes contain, on an average, 8,534 acres, and upwards of 1,200 fouls each.

‡ This is a little more than five to a houfe. The returns to the Revenue Board are at the rate of 5.59 to a houfe. *See Tranf. Royal Irifh Acad.* 1789.

4

for

for pasture. Few ditches are to be seen in this county, the fields being chiefly inclosed with dry stone walls*, which gives the country a dreary aspect.

The western part of the county is of quite a different character from the rest. The barony of Moycullin, which is also called Iarconnaught, contains some good land, on the sea coast and along the beautiful shore of *Lough Corrib.* But the heart of this barony is an assemblage of unreclaimable rock and mountain; and beyond *Oughterard,* Mount *Leam* stands very high above the lake. The rocks at *Oughterard,* and in the bed of the river *Fuogh,* of which there are immense masses, are all a black and white marble, equal at least in beauty with that of Kilkenny; yet there is seldom employment for more than one solitary artist, in working up a few chimney-pieces. *Lough Corrib* somewhat resembles *Lough Erne* in its form, and extends 20 miles in length, being 11 wide in the broadest part: in the middle it is contracted to a small channel, which is crossed by a ferry at *Knock* †. There is a fresh-water muscle in this lake that produces pearls; of which I have seen some very fine specimens. The large barony of Ballinahinch, which is better known by the name of Connamara, abounds with fine harbours, but is also extremely mountainous. The hills of *Ourred* and *Cashel* are very high, and the vast ridge called *Beannabeola,* or the *Twelve Pins,* which is a well-known sea-mark, consists of almost perpendicular rocks. At the foot of this ridge, close to the little village of *Ballinahinch,* a charming lake spreads itself for some miles; and on the river which runs from it into *Roundstone Bay* there is a great salmon fishery. On the sides of hills, and in the valleys, which are watered by rivers and small lakes, and sheltered in some places by the venerable remains of

---

* The same fences prevail in a great part of *Roscommon,* of *Mayo* and of *Clare.*

† A great number of concealed rocks render the navigation of this lake dangerous, to those who are not well acquainted with it.

N   ancient

ancient woods, the foil is moftly inclined to a black bog; but gravel, fand, or rock, lie at no greater depth than from one to three feet below the furface. Great quantities of *kelp* are made all along the coaft, and by manuring with fea wreck, the land is rendered very productive to the fcattered families that inhabit it, who are all little farmers, and hardy fifhermen. The northern part of Ballinahinch, and the barony of Ross, are called Joyces Country, and inhabited chiefly by a clan of that name. *Rofs* is alfo extremely rough ; *Mamtrafna*, on the borders of Mayo, is very high, and *Ben-Levagh*, at the north-weft angle of *Lough Corrib*, is a ftupendous mountain. Yet the borders of the lake, the fhore of the *Killeries*, and the valley through which the river *Bealnabrack* runs, are pretty well peopled, and the foil fuch as would amply repay the pains and expence of good cultivation.

This county, which reaches from the fea to the *Shannon*, is well watered by rivers and lakes : feveral of the rivers are, in part of their courfe, fubterraneous. The *Black-River*, on the bounds of Mayo, dips for about three miles near the village of Shrule. The *Clare* and the *Moyne* unite their waters underground, alternately appearing and retiring from view, in the *Turlachmore*; which in winter forms a lake, and in fummer a beautiful and found fheep-walk, upwards of fix miles in length, and two in breadth. Near Gort there are a vaft number of thefe *Swallows*; in which fome part of almoft every river and brook in the neighbourhood is ingulphed. The river *Gurtnamakin* dips feveral times, and after a concealed courfe of two miles, rifes on the beach below high-water mark, and difcharges itfelf among the rocks in the bay of *Kinvarra*. *Lough Rea* is a fine piece of water, and *Lough Coutra*, near the borders of *Clare*, is faid to poffefs all the beauties that hills, woods and iflands can impart to water.

The

The maritime advantages of this county muft not be omitted. The vaft bay of *Galway* is fheltered at the entrance by the *three* fouthern ifles of ARRAN*.   The found between thefe iflands is a fafe road ; and a number of inlets on the coaft, as well as the harbour of Galway, are fufficiently deep for the reception of merchant fhips ; but are more frequented by coafters and fifhing-boats than by veffels in the foreign trade.   The indented fhores of *Connamara* abound in well-fheltered havens—thofe of *Kill-kerran, Birterbuy, Roundftone,* and *Ballinakill,* are the largeft ; and the *Killeries* are at the northern extremity of this diftrict.

The town of GALWAY is fituated on the broad and ftony river, by which *Lough Corrib* empties itfelf into the fea.   It does not cover a very large fpace, but being very compact, and having little wafte ground within its ancient and mouldering walls, it con-tains a great number of inhabitants.   They may be eftimated at 12,000, though there are but 950 houfes ; for the greater part of this ancient town confifts of fquare edifices, at leaft two hundred years old, with each a fmall court in the centre.   Several diftinct families occupy thefe large houfes ; an arched way leading from the ftreet to the court, with a ftone ftair-cafe on each fide.   GAL-WAY was formerly the moft commercial town in Ireland ; but the fpirit of enterprife has long fince forfaken this once celebrated mart. The collegiate church, of which the conftitution and privileges are unique in Ireland, is very large.   There are three barracks in Galway, which are ufually garrifoned by two or three regi-ments of infantry.   TUAM, though an archiepifcopal fee, is but a very poor city.   LOUGHREA, on a lake of the fame name, is large and populous.   At ATHENRY, within an extenfive circuit of dilapidated walls, and their ruinous towers, the remains of caftles and abbies, that are intermixed with the cottages of a now

* Thefe iflands are very fruitful, and produce a fmall kind of oats, without any hufk. They are alfo remarkable for the flouted calves in the county.

fmall village, prefent a monument of its former confequence. It is remarkable, that old caftles are more frequent in this county than in any other part of Ireland. BALLINASLOE, on the weft fide of the river *Suck* (which is fo far navigable), though not fo large as fome others, is one of the moft thriving towns in the county ; and celebrated for a great wool fair in fummer, and a cattle fair in October, in which ten thoufand oxen and a hundred thoufand fheep are annually fold, from the paftures of *Clare* and *Mayo*, and of *this* great breeding County. The town of EYRECOURT, and the villages of DUNMORE, BALLIMOE, DUNAMON, HEADFORT, MONIVEA, CASTLE-BLAKENEY, &c. &c. are of little note.

Eight reprefentatives are deputed to Parliament by the *County*, the town of *Galway*, the city of *Tuam*, and the dilapidated borough of *Athenry*.

Obfervations on the Old Maps.

The omiffion of the *Killeries* has been already noticed: but the whole coaft of *Connemara* and *Iarconnaught* is inaccurate, and the names of thofe two diftricts confounded. The river *Beahnabrack* is made to flow into *Roundftone Bay*, and afford a fecond outlet to *Lough Corrib*, inftead of carrying into it a large fupply of water. In Jeffereys' map there appears no communication between the waters of the lake and the bay of Galway ; and *Oughterrard* is placed about ten miles N. W. of its true pofition. Indeed, the reprefentation of thefe three weftern baronies is quite erroneous in all the old maps. The ifles of *Arran* are placed too near the coaft of Clare, and mifcalled. *Inis-Bofin* is twelve miles from its true fituation. The boundary between the baronies of *Dunkillen* and *Loughrea* is incorrect; and the river which iffues from *Lough Rea* is made to run on the wrong fide of the town. The courfe of the *Moyne* is alfo inaccurate; and the *Turlach-more*, with another turlach near *Headford*, are

repre-

reprefented as permanent lakes. Many non-exifting villages are marked, even in *Connamara* and *Iarconnaught*, where there are only *Bunowen* and *Spiddal*, and thefe two are omitted—as are alfo, in the eaftern baronies, the villages of *Athlaggin*, *Mount-Bellew*, *Teinagh*, *Tobarfudder*, *Kinvarra*, and *Claran Bridge*. *Oranmore* is mifplaced *.

PROVINCE of MUNSTER.

MUNSTER comprifes *fix* counties, 61 baronies, and 816 parifhes. There are 3,377,150 acres, or 5,275 fquare miles †, and 184,546 houfes, in this province; which numbers, reduced to an average, will give 18.3 acres to a houfe, and 34.97 houfes in a fquare mile.

CLARE.

This County, which was anciently a part of MUNSTER under the name of *Thomond*, was added to CONNAUGHT in the reign of Elizabeth; but though it ftill continues in the Connaught circuit, it has long been reftored to the fouthern province. Bounded on

* Thefe weftern counties are fo little known, and yet fo interefting to the *Naturalift*, the *Philofopher*, and the *Legiflator*, that I have been tempted to dwell a little longer on their defcription than what the purpofed brevity of this Memoir would ftrictly allow. I muft at the fame time acknowledge, that with refpect to the two weftern baronies, although their outline is correct, and the face of the country truly reprefented in the new map, it is not pretended that the internal diftances and dimenfions are perfectly accurate.

† In Englifh meafure, 5,425,569 acres, or 8,474 fquare miles.

the eaſt and ſouth by the *Shannon*, and on the north-weſt by the *Atlantic Ocean*, it adjoins to the ſouth of Galway ; and extends from N. to S. 33, and from E. to W. 52 miles ; forming an area of 476,200 acres, or 744 ſquare miles *.

It is divided into 9 baronies, and 79 pariſhes +, and contains 19 churches. Of theſe, 57 pariſhes and 15 churches are in the dioceſe of *Killaloe* ; 19 pariſhes and 3 churches in the biſhopric of *Kilnefora* ; and 3 pariſhes, with *one* church, in that of *Lime-rick*. The baronies are, BURRIN, CORCOMROE, INCHIQUIN, IBRICKAN, MOYFERTA, CLANDERLAGH, ISLANDS, BUN-RATTY, and TULLAGH. They compriſe 17,396 houſes, and about 96,000 ſouls ‡. This population is at the rate of 27.3 acres to a houſe, and 23.38 houſes to a ſquare mile.

The baronies of *Burrin* and *Corcomroe* are exceedingly rocky ; but ſuch is the luxuriance of the paſturage interſperſed among the rocks, that theſe ſeemingly barren hills ſupport a great num-ber of cattle, and very large flocks of ſheep. There are many tracts of mountain in this county ; but the more level grounds are very fertile, and productive of corn and hay. In the rougher parts, a great number of excellent horſes are bred, which ren-ders *Ardſallis*, in the barony of *Burrin*, one of the principal horſe fairs in Ireland.

The river *Shannon* is from *one* to *five* miles broad between this county and that of *Kerry* ; and the *Fergus*, which is the principal river that riſes in Clare, forms a large eſtuary full of iſlands, at its junction with the *Shannon*. This river, and ſeveral others, dip underground in ſome part of their courſe. Here are alſo

---

* In Engliſh meaſure, 42 miles from N. to S. and 66 from E. to W. The area 765,042 acres, or 1195 ſquare miles.

+ The pariſhes contain, on average, about 6000 acres, and 1200 ſouls.

‡ At 5½ to a houſe, the proportion of inhabitants in this county appearing, by Mr. Buſhe's Tables, to be 6.4 to a houſe. *Tranſ. Roy. Ir. Acad.* 1789.

many

many turlachs: the moft remarkable is at *Kilcorncy*, in Burrin; where, as I am informed, the waters iffue, frequently more than once a year, from a fpacious cave, and deluge the adjacent flats.

ENNIS, the county town, is large and populous, and has the advantage of a fmall port at CLARE, which is fituated a few miles lower on the *Fergus*; the tide bringing up large boats from thence to *Ennis*. Except this town, and a confiderable fuburb to the city of *Limerick*, which ftands in the county of Clare, it can boaft no towns of note. CORROFIN, INISTYMON, SIXMILE BRIDGE, are fmall; as is alfo KILLALOE upon the Shannon, a place of great antiquity, and the fee of a bifhop. A great rock near this town, and a ledge of rocks lower down in the *Shannon*, impeding the navigation to Limerick, canals have been cut for the purpofe of getting boats and lighters up the river, which is navigable from hence to Lough Allen, with the interruption of a very few fhallows; at moft of which cuts * have been made. *Killaloe* is connected by a bridge with the village of *Ballina* in *Tipperary*. Juft above this town the *Shannon* contracts itfelf to the dimenfions of a river, after having, for 16 miles, expanded its waters to a very confiderable breadth, under the name of Lough *Deirgeart*.

The celebrated Ogham infcription, which was difcovered on the mountain of Callan in 1784 †, with feveral other Druidical remains, and the many rare plants which are produced in the mountainous and ftony parts of this county, render it equally interefting to the antiquary as to the botanift.

Four members of parliament are elected by the county of *Clare*, and the borough of *Ennis*.

* I omitted mentioning in its place, that a cut which was made at *Athlone* fome years ago, is become ufelefs by neglecting to keep the flood-gates in repair.

† See *Tranfactions of the Royal Irifh Academy for* 1787.

Obferva-

Obfervations on the Old Maps.

They extend the county 3 miles too much from eaſt to weſt. They repreſent the baronies of *Burrin*, *Corcomroe*, and *Clander-lagh* as flat, and they fill the firſt of theſe with imaginary villages; while that of *Curranroe-bridge*, on the borders of Galway, is omitted, as well as *Parteen*, in the neighbourhood of Limerick. The lakes and rivers of *Inchiquin* are incorrectly traced, and Lough *Ogram* is made a part of the Shannon.

LIMERICK.

Immediately ſouth of the Shannon, which parts it from Clare, the county of LIMERICK ſtretches from E. to W. 40 miles, being 25 broad from N. to S. It contains 386,750 acres, which make 604 ſquare miles *.

Excluſive of the *county of the city of* LIMERICK, and the *liber-ties* of KILMALLOCK, there are 9 baronies in this county— OWNEYBEG, CLANWILLIAM, COONAGH, SMALL COUNTY, COSHLEA, COSHMA, POBBLEBRIEN, KENRY, and CONNELLO, which laſt is almoſt, if not full as large as the other eight.

Theſe are divided into 125 † pariſhes, compriſing 33 churches; 84 pariſhes, with 24 churches, being in the diocese of *Limerick*; 38 pariſhes, and eight churches, in that of *Emly*; 2 pariſhes, with *one* church, in *Killaloe*; and one pariſh in the archbiſhop-rick of *Caſhel*.

---

* The length of Limerick, in Engliſh meaſure, is 51 miles, the breadth 32; and the ſuperficial contents 622,975 acres, or 970 ſquare miles.

† The average of the county is preciſely 3094 acres, and 1360 ſouls, to each pariſh.

This

This county is fo thickly inhabited, as to exceed moft of the counties of Leinfter in population ; for, eftimating it at the rate of 5½ * fouls to a houfe in the county, and 8½ in the city of Limerick, we fhall find 170,000 inhabitants, in the 28,748 houfes which this county contains : and this is in the average proportion of 47.59 houfes to a fquare mile, and of 13.4 acres to a houfe.

The foil of Limerick is extremely good for tillage, and very productive of grafs; efpecially thofe grounds which are called the *Corcachs* †, whofe fertility is proverbial, and is caufed by the rich manure which is annually depofited by the overflowings of the Shannon. The heavieft and fatteft beafts that are flaughtered at Cork are fed in this county ; much butter is exported from it; the orchards produce a very fine cyder, and it is by no means deftitute of trees and plantations.

This county, though diverfified by fmall hills, is not at all mountainous ; except on the fouth-eaft, where it is bounded by the *Galtees*, a ridge of formidable mountains that extends into Tipperary ; and on the borders of Kerry, where it grows uneven, and forms a grand amphitheatre of low but fteep mountains, which extends in a wide curve from *Loghil* to *Drumcolloher*. In the firft of thefe rifes the river *Maig*, which croffes the county and falls into the Shannon ; as do many fine ftreams, by which it is plentifully watered. In the weftern hills are the fources of the *Feale* and the *Gale*, which run weftward through Kerry ; and of the *Blackwater*, which flows in a contrary direction through the county of *Cork*.

The city of LIMERICK, a large diftrict, fituated in its peculiar *County*, is moft advantageoufly placed on the

* The abftract publifhed by Mr. Bufhe ftates the population to be at the rate of 6.11 to a houfe. *Tranf. Royal Irifh Academy*, 1789.

† *Corcach* fignifies, in Irifh, a fwampy ground, or marfh.

O

Shannon,

Shannon, which brings up to its keys ships of five hundred tons. The ancient part of the town is built in a large island which lies close to the eastern shore; and while it continued a fortified place, was esteemed the strongest in Ireland. It has been dismantled about 30 years, and has increased prodigiously within that period, by the addition of handsome streets and quays: its commerce has kept pace with its size, and great quantities of beef and other provisions are now exported from Limerick, which were formerly sent from this county to the port of Cork. As the number of houses in this city was 4866 in the year 1788, they may be fairly estimated at 4900 by this time, and the inhabitants at upwards of *forty thousand.*

There are several small towns and good villages in this county, of which RATHKEAL is generally esteemed the largest— KILMALLOCK was two centuries ago one of the best built inland towns in the kingdom. The walls of many large houses evidently of a date, of excellent workmanship, and all of cut stone, remain at this day; and the ruins of churches and monasteries, the walls and gates of the town, with the extensive district contained in the liberty of *Kilmallock,* prove the former splendour of a town, which is now no more than a miserable village. At CASTLE-CONNEL, on the Shannon, about 6 miles north of Limerick, there is a chalibeate spring, which has been found to possess the same qualities as the *Pouhon* at Spa. It is much frequented in summer, the situation of the village being delightful, and the accommodation tolerably good.

Eight members are deputed to parliament from the *County,* the city of *Limerick,* and the boroughs of *Askeaton* and *Kilmallock.*

Observa-

Obfervations on the Old Maps.

They add two miles to the length, and as many to the breadth of the county ; and are very incorrect in the courfe of the *Shannon*.—They omit the *County of the City*, and put *Limerick* in the barony of *Pobblebrien*; in which *Adair* alfo is improperly placed, the bounds of the baronies of *Cofhma* and *Kenry* being ftrangely confounded—*Kenry* they call *Kerry*.—They reprefent *Kilmallock* as a confiderable town, and they omit the villages of *Cullen*, *Caftletown*, *Any*, *Ardpatrick*, *Athlacca*, *Mungret*, *Killdeemo*, *Coolnakenny*, and *Nantinan*.

KERRY.

Weft of Limerick, and in the fouth-weft extremity of the ifland, the county of KERRY extends from N. to S. 53 miles, and in the broadeft part, from E. to W. 41 ; containing 647,650 acres, or 1012 fquare miles *.

It is divided into the eight baronies of IRAGHTICONNOR, CLANMAURICE, TRUAGHNACMY, CORCAGUINNY, MAGUNIHY, GLANEROUGHT, DUNKERRON, and IVERAGH, and contains 83 † parifhes and 20 churches; all in the diocefe of *Ardfert and Aghadoe*.

The population of this county is, after Galway, the thinneft in Ireland; the number of houfes being only 19,395, which is at the rate of 33.4 acres to a houfe, and of no more than 19.16

---

* In Englifh meafure, 67¼ from N. to S. 52 from E. to W. containing 1,040,487 acres, or 1639 fquare miles.

† The parifhes contain, on an average, 7800 acres, and 1280 fouls.

houfes

houfes in a fquare mile. The number of inhabitants may be about 107,000, if we allow $5\frac{1}{2}$ to a houfe *.

It is not furprifing, that this county fhould be thinly inhabited : barren mountains, and almoft inacceffible rocks, render a large portion of it unfit for habitation, and incapable of culture. Even the northern baronies, in which there is much good land, with few mountains, are far from level ; and the chearful afpect of cultivated fields and fine paftures is frequently interrupted by bleak and ftony hills ; while tracts of bog interfect the narrow plain that extends from Caftlemain harbour to the borders of the county of Cork †, between the *Mang* and the *Flefk.* Grazing is more attended to than tillage, and this part of *Kerry* fupplies many fat beafts of good fize, and great numbers of ftore cattle : but the native breed of the country is extremely fmall, yet remarkably good for the pail, refembling the Alderney cow both in fize and character ; butter is confequently a confiderable article among the exports of Kerry. The barony of *Corca-guinny* forms a peninfula between the bays of Dingle and Tralee, and terminates at *Dunmore Head,* the moft weftern point of Ireland and of Europe. It is full of mountains, but the high promontory of *St. Brandon* is eminent above the reft, and the mountain of *Cahirconree* ftands acrofs the ifthmus. Among the rough and high hills in the barony of *Iveragh,* and the fouthern part of *Dunkerron,* fome pleafant vallies and improveable grounds are interfperfed ; and in the ifland of *Valentia* there are more inhabitants, and a better culture, than could be expected in fo remote a fpot. *Glanerought* is entirely covered with exceeding high and rugged hills, and feparated from the county of

* The population in Mr. Bufhe's Paper is returned at the rate of 6.29 to a houfe. *See Tranf. of Roy. Irifh Acad.* 1789.

† This plain is continued through the whole length of *Cork,* till it is clofed by the mountains of *Knocknoledown,* and the *Galtees.*

Cork

Cork by an immenfe and almoft impaffable ridge of rocky mountain; over which there is but one pafs, and that very difficult, called
the *Prieft's-leap:* but the loftieft mountains in this county ftand
in a huge affemblage on the weft and fouth of *Killarney,* half
encompaffing the lower, and entirely furrounding the upper lake.
Of thefe, *Mangerton* is generally efteemed the higheft, being
2500 feet above the fea; but it is doubted, whether the craggy
fummits of *Macgillycuddy's Reeks* do not furpafs it in altitude. In
this defolate tract there are large herds of red deer, and abundance
of game.

Of all the Irifh lakes, *Lough Lane,* near the town of *Killarney,*
is defervedly the moft celebrated for picturefque beauty. In the
lower and larger lake, the pleafing and the fublime are moft happily combined; the upper lake reflects a more folemn grandeur
from the ftupendous crags with which it is encircled: but the
profpects in both are infinitely diverfified. The feveral iflands,
the white rocks of Mucrufs, the groves of Arbutus, the venerable woods, the variety of waterfalls, and the impending cliffs—
are feparately as delightful and interefting, as their affemblage is
eminently grand and magnificent. There is alfo a fmall lake in
Glancrought, about ten miles from Nedeen, called *Lough Cloney,*
which is faid to poffefs all the charms of Killarney in miniature.
Many fine rivers water this county.—The *Cafhin,* which is formed
by the union of the *Feale* and the *Gale,* is navigable for eight or
ten miles—The *Lane* flows out of the lake of Killarney, which
receives the *Flefk*—The *Roughy* pours its impetuous current into
Kenmare river—The *Mang,* which is navigable to *Caftlemain,* was
the northern boundary of the ancient County-Palatine of *Def-
mond,* and falls into *Caftlemain* harbour, at the bottom of the
great *Bay of Dingle,* which can only admit veffels of moderate
burden.-- *Ventry Bay,* the roads of *Dingle* and *Valentia,* and *Bal-
lynafkelligs Bay* are fmall but commodious harbours. *Kenmare*
                                                              River

River is a frith or eftuary, which extends twenty miles in length, and widens to a breadth of three, and is one of the moft fecure and capacious havens in Europe; but it has been little frequented fince the Pilchards, for which there was formerly a fuccefsful fifhery, have forfaken that coaft. NEDEEN, or KENMARE TOWN, notwithftanding its cotton manufactory, and its advantageous fituation, at the mouth of the *Roughy*, is but a poor village.

ARDFERT, or ARDART, is alfo very fmall; though, by being the fee of a bifhop, and fending members to parliament, it comes within the definition of a *city*. The round tower, which had ftood there for ages, fell a few years ago, tumbling at one crafh into a heap of ruins.

TRALEE, the county town, is pretty large—Near it there is a chalibeate fpring, which is drank medicinally, with good fuccefs. KILLARNEY is fmaller, but in a very improving ftate; not fo much owing to the great concourfe of vifitants, whom the beauties of the lake attract, as to the judicious attention of the refpectable proprietor, Lord Kenmare, and to the encouragement which he gives to the extenfion of the linen manufacture through that neighbourhood. MILLTOWN, which was a fmall village but a few years fince, promifes to become a good market-town by the exertions of Sir William Godfrey, and the convenience of water carriage; the tide bringing up floops from the Mang very near this town. DINGLE, or *Dingle-i-couch*, was once of great note and of good commerce; and though the town fhews at prefent fome evident marks of decay, yet it ftill preferves fome little trade, and exports butter, beef, corn, and even linen.

Near this town, in the caves on the fea fhore, there is an abundance of very clear and hard cryftals: and fome amethyfts of great beauty and luftre are found about *Kerry Head*— Pearls have alfo been taken out of the lake of Killarney. There

5   are

are in several parts of the county marble and slate quarries; and it is not destitute of coal-mines, but the abundance and cheapness of turf or peat, render them useless. Several mines of copper, lead, and iron, were formerly worked to good effect, and furnaces were established, while timber was plenty; but now that the country is completely cleared of wood, they are neglected. The cider of Kerry, which is made of the cackigay apple, is highly prized, and brings a great price; yet orchards are not very numerous, and that valuable fruit, with another excellent apple, the Kerry pippin, are little propagated, and difficult to be procured.

Eight members are deputed to parliament by the county of *Kerry*, by *Tralee*, *Ardfert*, and *Dingle*.

Observations on the Old Maps.

They make the County one mile shorter from N. to S. and *eight* miles broader from E. to W. than the new: but, as a proof that the reduced breadth in the new map is more correct than the old, the area of the County, as measured upon the new map, exceeds the dimensions usually attributed to it * by upwards of *ten thousand* acres. Those maps represent the harbour of *Castlemain* too large, and are incorrect as to both the form and position of the *Blasket* or *Ferriter* Islands, and of the *Skeligs*. They omit the villages of *Galey*, *Killurry*, *Smerwick*, and *Ti-vourney-geraan*, near *Dunmore Head*, of as much celebrity in the west of Ireland as John a-Groats house in the north of Scotland.

* See Watson's Almanack.

CORK.

CORK.

Eaſtward from Kerry, the county of Cork ſtretches 78 miles, extending from N. to S. 56. It is the largeſt county in Ireland; containing 1,048,800 acres, or 1638 ſquare miles *.

It comprehends 16 baronies—DUHALLOW, ORRERY AND KILMORE, FERMOY, CONDONS AND CLANGIBBON, KILLNA-TALLOON, IMOKILLY, BARRYMORE, BARRETS, MUSKERRY, KINALMEAKY, KINALEA AND KERRYCURRIHY, COURCEYS, BARRYROE, IBAWNE, CARBERY, and BEAR AND BANTRY; to which muſt be added *four* peculiar diſtricts. The *county of the city of* CORK, and the *liberties* of YOUGHAL, KINSALE, and MALLOW.

This county is divided into † 269 pariſhes, in which there are 105 churches; the dioceſe of *Cork*, compriſing 94 pariſhes and 41 churches, and that of *Roſs* 33 pariſhes and 12 churches. The biſhopric of Cloyne contains 137 pariſhes and 51 churches; and 5 pariſhes with *one* church, belongs to the ſee of *Ardfert*.

The population of this county is confiderable, it contains 76,739 ‡ houſes, of which about 8100 § are in the city of Cork. The average of the whole gives 46.85 houſes to a ſquare mile, and 13.65 acres to a houſe. The number of inhabitants in the city muſt be near 73,000, and in the reſt of the county about 343,000, making in all 416,000 ſouls ‖.

---

* The county of Cork is much larger than any county in England, except Yorkſhire. Its length being, in Engliſh meaſure, 99 miles; its breadth 71½; and its area 1,67,920 acres, or 2,653 ſquare miles.

† The extent of theſe pariſhes would be, on an average, about 3900 acres, and their population 1550 ſouls.

‡ This was the number at Chriſtmas 1791, as returned to parliament, while theſe ſheets were printing.

§ At the end of 1788 the houſes in the city amounted to 8073.

‖ The population of this county was found to be at the rate of 5.6 to a houſe, and that of the city 9.06. *Tranſ. Royal Iriſh Acad.* 1789.

tains,

In fo large a diftrict there muft be a great variety of foil. It contains more good land than bad, and fome parts of the county are highly cultivated.

The barony of *Bear and Bantry* which is covered with mountains, and the weftern parts of *Carberry* and *Mufkerry*, in which are the *Sheehy* mountains, are the pooreft and the leaft improved.

The whole county is hilly, and, a few places excepted, very deftitute of trees. The *Galtees* and the *Waterford* mountains bound it on the north-weft. The *Nagle* mountains and the *Bogra*, which run weftward through the heart of the county, are a part of that range which is continued, with few interruptions, from Helwick Head in Waterford, acrofs the counties of Cork and Kerry to the ocean; and on the north of this ridge lies the narrow plain, which extends from the bounds of Tipperary to Dinglebay.

The county of Cork abounds in fine rivers and good harbours. The *Blackwater* rifes in the mountains between Limerick and Kerry, and flows eaftward through this county, receiving the *Funcheon*, the *Bride*, and many fmaller ftreams, in a courfe of eighty miles. It is navigable to *Cappoquin*, in Waterford, where it turns to the fouthward, and difembogues itfelf in the harbour of Youghall. The *Lee* iffues from a lake on the weft of *Inchigeela*, and paffing through the city of Cork, in a broad and deep channel, contributes efpecially to the wealth and profperity of that great commercial town. The *Bandon* is another fine river, which after watering the large and thriving town of *Bandon Bridge*, and the neat village of *Inifhonan*, falls into the harbour of *Kinfale*: it is navigable for large floops as far as *Inifhonan*, between the moft beautiful wooded and winding banks. The whole coaft of CORK is broken into creeks and bays. The town and harbour of CROOK-HAVEN, near *Mizen-head*, the fouth-weft point of Ireland, are very well known. *Bantry-bay* which lies a little to the north of

P   this

this cape is at leaſt twenty miles long, and from three to five broad; every where deep, ſheltered, and free from rocks. But the town of Bantry, though at the bottom of this noble bay, is little benefited by its ſituation, ſince the pilchard fiſhery * is now loſt. Ships of war and merchantmen often put into *Bear-haven*, near the mouth of this bay; but the village of *Caſtletown* affords them few reſources, and no accommodation. At *Adragool*, in this bay, there is a cataract of prodigious height, which is diſtinctly ſeen from Bantry, above ten miles acroſs the bay.

Kinsale is well known to poſſeſs an excellent and important harbour, though a bar prevents large men of war from coming into the baſon cloſe to the town; in this port alone there is a dock with ſtores for the uſe of king's ſhips. The entrance of the harbour is defended by a fort, which, having been conſtructed in the reign of Charles II. is called *Charlesfort*, and in which there is always a good garriſon. The town which contains at leaſt 10,000 + inhabitants is built on the ſide of *Compas-hill*, and cloſe to the water's edge. The broad ſummit of this hill would be the moſt commanding and advantageous ſituation in Ireland, for a fortified town or citadel.

The harbour of Cork is ſo formed, as to contain an immenſe number of ſhips, in complete ſecurity. And the banks of it, being adorned with villas and plantations, preſent a moſt agreeable and chearful landſcape. Veſſels of 120 tons go up to the city quays, but the large ſhips lie at *Paſſage*, a few miles lower down: the mouth of the harbour is protected by *Carliſle* fort, the date of whoſe conſtruction is pointed out by the name. This city is the great mart of the South of Ireland. The principal articles of its export trade are beef and butter: upwards of *fifty*

---

* While I was at Bantry, in 1788, ſome pilchards were taken, from which ſanguine hopes were entertained of the return of that fiſh to the Iriſh coaſt; but I have not ſince heard that theſe hopes have been realized.

+ The number of houſes in 1783 was 1079.

thouſand

thoufand barrels of the former, and about *feven thoufand* tons of butter were exported in the courfe of laft year; they export alfo a great quantity of corn, and fome linen. *Youghal* is fituated at the eaftern extremity of the county, on the mouth of the *Black-water*, and contains about 7000 inhabitants ; among whom feveral genteel families are permanent, exclufive of a great refort of bathers in the fummer feafon. The port is commonly full of veffels engaged in the coafting trade.

Of the inland towns, MALLOW is the moft confiderable; it is fituated on the fame river, forty miles higher up, and very much frequented, on account of a foft and tepid fpring, of the fame nature and efficacy as the Hotwells of Briftol.

CLOYNE, though the refidence of the bifhop, is of little confequence; and Ross is ftill fmaller—But to enumerate all the towns and principal villages of this extenfive county would exceed my bounds. MITCHELSTOWN however claims a particular notice, both for the elegance and regularity of its buildings, and on account of a college, founded a few years ago by the late lord Kingfton, for twenty-eight decayed *gentlemen* or *ladies*, who receive 40 l. a year each, and have a neat houfe and garden. The college is a handfome range of regular buildings, with a chapel in the centre, forming one fide of a large fquare. There is alfo a good houfe, with a falary of 100 l. a year for the chaplain.

At *Dromagh* and *Dromanagh*, in the barony of *Duhallow*, there are coal-pits; iron is alfo raifed, and there are fome furnaces in this county. Much linen is wove about the *Inifhonan* and *Dunmanaway*; cotton and other manufactures have been eftablifhed at *Blarney*, and coarfe woollens are made in feveral parts of the county. Of the many iflands appertaining to this county, the moft noted is *Clare-ifland*, the fouthern point of which is well known to mariners by the name of CAPE CLEAR, but which ought rather to be written Cape *Clare*.

This county and its towns are reprefented in parliament by 26 members, who are delegated by the *county*, and the *city* of CORK,

the

the towns *Kinfale, Youghal, Bandon-bridge, Mallow, Doneraile, Rathcormuck, Middleton, Charleville,* and *Caftlemartyr,* and by the fmall boroughs of *Baltimore* and *Clonekilty.*

Obfervations on old Maps.

They make the county five miles too long from E. to W. and two miles too broad from N. to S. placing *Cork* city, as has been already noticed, 8 minutes and 54 feconds *fouth* of its true latitude, and feven minutes and a half of longitude *too near* the meridian of Dublin. The divifions of baronies are inaccurate. *Bear Ifland* is quite too far from the fhore: The form of *Clare Ifland* is erroneous, and it is placed too near the coaft. They omit the villages of *Glanton, Ballyclough, Ballydonnel, Caftle Hyde, Knockmourney, Conno, Ballyvourney, Caftletown, Ballykilly, Caftle Townfend, Douglas, Monkftown, Shangan,* &c.

WATERFORD.

This maritime county, which is fituated at the eaft of Cork, extends from E. to W. 40 miles, and from N. to S. 23; and contains 262,800 acres, or 410 fquare miles *.

It comprifes the *liberty* or *county of the city* of WATERFORD, and the *feven* baronies of COSHMORE AND COSHBRIDE, DECIES *within* DRUM, DECIES *without Drum,* GLANEHIRY, UPPER THIRD, MIDDLE THIRD, and GUALTIERE: which contain 74 † parifhes and 21 churches; 34 parifhes and 8 churches lying in

---

* The dimenfions of Waterford, in Englifh meafure, are—length from E. to W. 51 miles, breadth from N. to S. 29, area 425,692 acres, or 665 fquare miles.

† The average of each parifh is above 3500 acres, and about 1400 fouls.

the

the bifhopric of *Waterford*; and 40 parifhes with 13 churches in the diocefe of *Lifmore*. There are 18,796 houfes, and at leaft 110,000 * inhabitants in the county of Waterford. And the average of the whole county, including the city, is at the rate of 45.84 houfes to a fquare mile, and 13.98 acres to a houfe +. Very little of this county is level; but in the fouth and eaft, the foil though hilly is rich and productive. A group of mountains overfpreads a confiderable fpace between Dungarvon to Clonmell, of which the higheft range is called the *Commeragh*. Another ridge extends on the north of the Blackwater, to the borders of Cork and Tipperary, under the name of *Knockmeledown*. Thefe hills, however, except in a few defolate and craggy fpots, afford pafture to fmall cows, which produce a great quantity of butter. Dairies are alfo very numerous in the more level grounds of of the county.

The river *Blackwater* flows through the weft of this county, into the bay of Youghal; and is navigable to *Cappoquin*. The gentle and majeftic *Suir* divides it from the counties of Tipperary and Kilkenny, running eaft till joined by the *Barrow*; when turning to the fouth, they form an eftuary nine miles long and two broad, which is the harbour of Waterford.

The large and populous city of WATERFORD is well built, and contains about 35,000 fouls. It ftands on the fouth fide of the river *Suir*; which is embanked by a very noble quay, extending the whole length of the town. Veffels of great burthen can come up to this quay, but the largeft fhips generally lie a few miles lower down. A very flourifhing commerce, with England and other countries, is the happy confequence of fuch a fituation.

* At 5¼ to a houfe in the county, and 9 in the city. By Mr. Bufhe's returns they were found to be 6.30 to a houfe in the county. *Tranfa7. Royal Irifh Academy*, 1789.

+ In the year 1788 the city contained 4097 houfes, according to Mr. Bufhe's paper. Ibid.

The

The principal articles of export are beef, pork, butter, grain and linen.   Packet-boats are also established between this port and Milford-haven, for the convenience of the south of Ireland.

As there is no bridge across the Suir below *Carrick*, the only communication with Leinster is by a ferry, an inconvenience to which the great depth and breadth of the river has hitherto compelled the inhabitants to submit.   But it is hoped, that they will soon evince the same spirit as the citizens of Derry, and follow their example, with equal success.

At the small village of P A S S A G E, outward bound ships usually wait for a wind.   Near it stands the N E W - G E N E V A, an elegant village regularly built, which was erected by government a few years since, for the reception of the expatriated citizens of Geneva. But they having relinquished the design of settling in Ireland, this place remains still uninhabited.   Almost opposite, on the Wexford shore, the fort of D U N C A N N O N protects the harbour of Waterford, by batteries of heavy cannon ; the deep part of the channel running close to the rock on which the fort is built. *Hooktower*, at the extremity of the narrow peninsula on the E. of the harbour, is used as a light-house.   T R A M O R E, a village six miles S. of Waterford, consists of a number of neat houses, situated on a fine strand, at the edge of a shallow bay, and much resorted to for sea-bathing.

D U N G A R V A N is a good fishing-town on a small harbour, and enjoys a considerable share of the coasting trade.   It is supplied with fresh water from the river *Phinix*, by an aqueduct of about six miles in length.   T A L L O W is a thriving town, the river *Bride*, which passes within half a mile of it, and falls into the *Blackwater*, being so far navigable for large boats.   On the southern bank of the *Blackwater* stands L I S M O R E, in early times a considerable city, now but a small and dilapidated town.   Here a noble bridge of one arch, *ninety* feet in the span, has been lately

thrown

thrown acrofs the river.   For many miles round, the roads are lined with apple-trees, and the country covered with orchards. CAPPOQUIN and KILLMACTHOMAS are good villages.

The *county* and *city* of Waterford, and the boroughs of *Lifmore*, *Tallow*, and *Dungarvan*, return *ten* members to parliament.

Obfervations on the Old Maps.

They add one mile to the length, and three miles to the breadth of the county.   *Tallow* is placed *on* the Bride, and *Lifmore* at fome diftance from the Blackwater.   In fome maps the mountains of *Knockmeledown* are entirely omitted, and in Jefferys' the villages of *Villierflown*, *Tallowbridge*, *Rockville*, *Whitechurch*, *Roffmore*, and *New Geneva*.

TIPPERARY.

THE county of TIPPERARY * joins that of Waterford, and ftretches northward 52 miles, terminating like a wedge, between Leinfter and Connaught.   From E. to W. it meafures 31 miles, and contains 554,950 acres, or 867 fquare miles †.

TIPPERARY comprehends the 12 baronies of IFFA and OFFA, CLANWILLIAM, MIDDLETHIRD, SLEWARDAGH and COMPSEY, KILNEMANNA, KILLNALONGURTY, ELIOGURTY, IKERIN, ILEAGH, OWNEY AND ARRA, UPPER ORMOND, and LOWER ORMOND ; which are divided into 186 ‡ parifhes, and comprife 46 churches.   Of thefe 94 parifhes and 22 churches,

---

* The greateft part of this county was a *Palatinate* in the Ormond family, from 1328 to 1716, when the jurifdiction was abolifhed by act of parliament.

† In Englifh meafure this county is 73¼ miles long and 39½ broad : and contains 882,398 acres or 1420 fquare miles.

‡ The parifhes contain, on an average, about 3000 acres, and 900 fouls.

are

are in the archbifhoprick of *Cafhel*, and 20 parifhes with 4 churches in the bifhoprick of *Emly*. To the fee of *Lifmore* belong 31 parifhes and 8 churches, and to that of *Killaloe* 41 parifhes with 12 churches.

There may be about 169,000 inhabitants in this county, eftimating the population of 30,703 houfes at 5½ fouls * per houfe. And the average proportion of houfes is 18.07 acres to a houfe, and 35.51 houfes in a fquare mile. This is about the medium population of the whole province, but very fcanty for fo fine a county. For except in the rough hills of *Kilnemanna* and *Owney*, fome mountains near *Rofcrea*, the lofty *Keeper*, *Sliebh-na-Man*, the *Galtees* and *Knockmeledown*, all of which occupy but a fmall proportion of this extenfive county, the foil is generally very good, and in fome large tracts, particularly in that which is called the *Golden Vale*, and about the town of *Tipperary*, extremely rich and fertile. But this county is more celebrated for the excellence of its cattle, and the verdure of its fheep walks, than for the number of its corn fields. Much wheat however is raifed, chiefly in the fouth, fince it has no lefs than 48 boulting mills, which is a greater number than any other county can boaft.

Small rivers and brooks in abundance fupply thefe mills: and the *Suir* rolls through the heart of the county. This river rifes in the mountain of *Bendubb*, on the borders of the King's County, and takes a fouthern direction, till, obftructed by the Waterford mountains, it is compelled to alter its courfe; when, turning eaftward, it flows in a deep and broad current, at the foot of thefe hills, paffes under the bridges of *Clonmell* and *Carrick*, and after uniting with the *Nore* and the *Barrow*, meets the fea, near a hundred miles from its fource.

† It appears from Mr. Bufhe's paper, that as far as it was inveftigated the inhabitants are found to be at the rate of 6.20. See *Tranfact. of Royal Irifh Acad.* 1789.

I   CLONMELL.

CLONMELL is the fhire town, large and opulent, where the woollen and cotton manufactures are in a flourifhing ftate. Though very inconveniently fituated for the affizes, at the extremity of fo large a county; it is admirably feated for trade, on the northern bank of the *Suir*, which is fo far navigable for large boats, the tide flowing a little way above the town. Sloops of confiderable burthen reach CARRICK, which is alfo a good town, famous for the manufacture of a particular kind of woollen cloth, called *Ratteen*. The city of CASHEL is well inhabited for its fize, but has no trade. FETHARD, TIPPERARY, CAHIR, HOLY-CROSS, THURLES, SILVERMINES, NENAGH, and ROSCREA, are the principal towns.

The leadworks at *Silvermines* are very productive, and fome virgin filver has been found among the ore ; but the chief products of the county are butter, fat cattle, fheep, and flour, of which laft article great quantities are fent to Dublin.

The county of *Tipperary*, the city of *Cafhel*, *Clonmell*, and *Fethard*, are reprefented in parliament by 8 members.

Obfervations on the Old Maps.

In them the county meafures 54 miles in length, and 36 in breadth. The high mountains of *Bendubh* and *Sliebh-na-Man* are omitted. *Knockmeledown*, and the *Galtees* are ill expreffed. The courfe of the *Suir* is not correct ; and fome part of the boundary with Limerick is erroneous. In Jefferys' map the following villages are not marked, *Modreeny*, *Ballina*, *Newport*, *Burros-Ileagh*, *Templemore*, *Ballybeg*, *Lickflin*, *Mullinahone*, *Golden*, and *Ardfinnan*.

Q   ECCLE-

III. ECCLESIASTICAL DIVISION OF IRELAND.

THE ecclefiaftical ftate of this kingdom is ftill lefs known than its topography, nothing authentic having yet appeared in print upon that fubject. The ftatements in the following pages are founded on the authority of the regiftries and vifitation books of the refp ctive diocefes; on the communications with which the author has been favored by feveral of the bifhops and clergy; and on the information which he acquired in vifiting the different parts of the kingdom.

The firft preachers of chriftianity in Ireland eftablifhed a great number of bifhopricks, which gradually coalefced into the *thirty-two* diocefes, that have for feveral centuries conftituted the ecclefiaftical divifion of the kingdom. But when the country became impoverifhed and depopulated, by the perpetual feuds and frequent civil wars with which it was defolated for ages; it was found neceffary at different periods to unite fome of the pooreft of thefe fees, in order that the bifhops might have a competence to fupport the dignity and hofpitality incumbent on their high ftation: and hence it comes, that there are only *twenty-two* prelates in the church of Ireland, *twenty* fees being united under *ten* bifhops. Thefe caufes having had the fame operation with refpect to parifhes, the 2438 parifhes do not form quite 1200 benefices; many having been confolidated by the privy council, from time to time, under the authority of an act of parliament; and many others, though but epifcopally united, having been confidered as only one living time out of mind.

This

This kingdom is divided ecclefiaftically, as well civilly, into *four* provinces ; but the civil and ecclefiaftical boundaries are far from coinciding. An archbifhop prefides over each. The *feven* bifhops of the northern province are fuffragans to the archbifhop of AR-MAGH, who is LORD PRIMATE, and *metropolitan of all Ireland.* The archbifhop of DUBLIN is *Lord Primate of Ireland,* and has *three* fuffragan bifhops in the eaftern province. The fouthern province with its *five* fuffragans is under the jurifdiction of the archbifhop of CASHEL, *Lord Primate of Munfter.* And the arch-bifhop of TUAM, *Lord Primate of Connaught,* prefides over the *three* bifhops of the weftern province.

The number of Deanries in this kingdom is *thirty-three,* and of Archdeaconries *thirty-four.* But the archdeacons have not a vi-fitatorial jurifdiction ; the government of the church of Ireland, which is in moft things conformable to that of England, differing with refpect to vifitations: for in Ireland, the bifhops hold a vifita-tion annually, and the archbifhop vifits his fuffragans every third year.

In defcribing the prefent ftate of the feveral diocefes, their prin-cipal circumftances are, for the greater precifion and concifenefs, reduced to tables ; which being divided into *ten* columns, exhibit in the

1ft. The COUNTIES into which the diocefe extends.

2d. The number of ACRES which it includes.

3d. The grofs number of PARISHES in each.

4th. The number of BENEFICES into which thofe parifhes are at prefent *united.*

5th. The number of CHURCHES.

6th. The number of GLEBE-HOUSES.

7th. The number of parifhes which have GLEBES *without houfes.*

8th. The number of *benefices* which have NO *glebes.*

Q 2   9th.

9th. The number of vicarages, the *rectories* of which are lay IMPROPRIATIONS.

10th. The number of parishes which are WHOLLY IM-PROPRIATE.

Whenever parishes lie in more than one county, they are numbered to that county in which the church is fituated.

THE PROVINCE OF ARMAGH CONTAINS * TEN DIOCESES.

ABPK. of ARMAGH	*Bpk.* of Raphoe
Bpk. of Dromore	—— of Clogher
—— of Down ⎱ *united.*	—— of Kilmore
—— of Connor ⎰	—— of Ardagh †
—— of Derry	—— of Meath

THE PROVINCE OF DUBLIN CONTAINS FIVE DIOCESES.

APK. of DUBLIN	*Bpk.* of Ferns ⎱ *united.*
Bpk. of Kildare	—— of Leighlin ⎰
—— of Offory	

THE PROVINCE OF CASHEL CONTAINS ELEVEN DIOCESES.

APK. of CASHEL ⎱ *united.*	*Bpk.* of Cloyne
Bpk. of Emly ⎰	—— of Limerick ⎤
—— of Waterford ⎱ *united.*	—— of Ardfert ⎬ *united.*
—— of Lifmon ⎰	and Aghadoe ⎦
—— of Cork ⎱ *united.*	*Bpk.* of Killaloe ⎱ *united.*
—— of Rofs ⎰	—— of Kilfenora ⎰

THE PROVINCE OF TUAM CONTAINS SIX DIOCESES.

APK. of TUAM	*Bpk.* of Elphin
Bpk. of Clonfert ⎱ *united.*	—— of Killalla
—— of Killmacduagh ⎰	—— of Achonry

---

* The diocefes are placed with refpect to contiguity, not according to rank ; for the bifhop of *Meath* has precedence of all bifhops, and next to him *Kildare* ; the other bifhops take place according to the date of their confecration.

† *Ardagh*, though in this province, is at prefent annexed to the archbifhoprick of *Tuam*.

PROVINCE OF ARMAGH.

1. ARCHBISHOPRICK OF ARMAGH.

THIS fee was founded by St. Patrick about the middle of the fifth century, and was made an *archbishoprick* in the year 1152. It extends into *five* counties, 59 miles from N. to S. varying in breadth from 10 to 25 *.

Counties.	Acres.	Parishes.	Benefices.	Churches.	Glebe Houses.	Glebes only.	Benefices without glebe.	Rectories impropr.	Vicar'y impropr
Armagh -	170,850	17	17	23	23†	1	1	–	–
Londonderry	25,000	5	5	6	4	1	–	–	–
Tyrone -	162,500	20	19	20	13	6	–	–	–
Louth -	108.900	61	28	20	11	5	13	12	9
Meath -	1,300	part of two	–	–	–	–	–	–	–
Total	468,550	103	69	69	51	13	14	12	9

The CHAPTER consists of a Dean, Precentor, Chancellor, Treasurer, Archdeacon, and 4 Prebendaries, with 8 Vicars-choral.

PATRONAGE. The Crown has the presentation to 13 parishes, the lord Primate to 60, the university to 5, and the chapters of Christchurch and St. Patrick Dublin to 3, the remainder have lay patrons.

In the city of Armagh, which is 35 miles distant from the extremity of the diocese, there is a cathedral with a good choir; and a very handsome archiepiscopal palace has been erected by the present lord Primate.

* In English measure this diocese is 75 miles long, and from 12½ to 32 broad.

† Four of these Glebe-houses are on the perpetual cures into which the parish of Armagh is divided, and there are five more appropriated to the choir.

3

2. BISHOP-

2. BISHOPRICK of DROMORE.

The foundation of this diocefe is afcribed to St. Colman in the 6th century. It is extremely compact, and the fmalleft in extent of any bifhoprick in the kingdom, which is not annexed to another fee; extending only 28 miles from N. to S. and 17 from E. to W. * Yet it comprehends fome part of three counties.

Counties.	Acres.	Parifhes.	Benefices.	Churches.	Glebe Houfes.	Glebes only.	Benefices without glebe.	Rectories improp.	Wholly improp.
Armagh	10,600	3	3	3	2	–	1	–	—
Down	143,700	22	20	23	12	2	6	2	—
Antrim	1,500	1	1	1	–	–	1	1	—
Total	155,800	26	24	27	14	2	8	3	—

The CHAPTER of this diocefe, which was new modeled and eftablifhed, with fome peculiar privileges, by patent of James I. is compofed of a Dean, Precentor, Chancellor, Treafurer, Archdeacon and one Prebendary.

PATRONAGE—The deanry alone is in the gift of the Crown †; *one* parifh is in the lord Primate; 23 in the bifhop of Dromore, and 2 are in laymen.

The lordfhip of NEWRY claims the fame exemption from epifcopal jurifdiction, to which it was entitled when it appertained to a monaftery, before the reformation.   And the pro-

* The extent of Dromore, in Englifh meafure, is 35½ by 21½.

† The patronage of the deanry is conceded to the bifhop by King James's patent, but the Crown has continued to prefent.

prietor

prietor of the lordſhip (Mr. Needham) exerciſes the juriſ-
diction in his peculiar court, granting marriage licenſes, probates
to wills, &c. under the old monkiſh ſeal.

The cathedral of Dromore is very ſmall, but the biſhop's houſe
which was erected a few years ago by Dr. Beresford, the preſent
biſhop of Oſſory, is a handſome and convenient reſidence, near
the town; and not twenty miles diſtant from any part of the
dioceſe.

---

3 and 4. BISHOPRICKS of DOWN and CONNOR.

Theſe biſhopricks were both founded in the 5th century, and
united in the year 1454. The greateſt length of DOWN is 41
miles from N. E. to S. W. and the greateſt breadth 22 *.
CONNOR extends in length N. and S. 45 miles, and in breadth
24 †. The extent of the united ſees, from the north of Antrim
to the ſouthermoſt point of Down, is *ſeventy* ‡ miles.

There is part of one pariſh of the dioceſe of DOWN, in the
county of *Antrim*, the remainder are all in the county of *Down*.
The ſee of CONNOR lies chiefly at *Antrim*, but includes alſo a
part of *Down* and *Londonderry*.

* In Engliſh miles 52 and 28.
† Length 57, breadth 30½, Engliſh miles.
‡ The length of the union 89.

DOWN.

D O W N.

County.	Acres.	Parishes.	Benefices.	Churches.	Glebe Houses.	Glebes only.	Benefices without glebe.	Rectories improp.	Wholly improp.
Down - -	201,150	38	25	33	12	5	9	3	5
Antrim - -	800	Part	—	—	—	-	-	-	-
Total	201,950	38	25	33	12	5	9	3	5

C O N N O R.

Down	3,700	part of one	—	—	—	—	—	—	—
Antrim	382,400	73	39	41	10	11	17	14	10
Londonderry	9,400	3	1	2	1	1	—	—	—
Total	395,500	76	40	43	11	12	17	14	10
Total of union	597,450	114	65	76	23	17	26	17	15

The CHAPTER of thefe Sees was alfo regulated by patent of James I. That of DOWN confifts of a Dean, Precentor, Chancellor, Archdeacon, and two Prebendaries. The fame dignitaries are in CONNOR, with 4 Prebendaries.

The fame exemption is claimed, by the Needham family, for their lordfhip of Mourne in this diocefe, as for that of Newry in Dromore; but the claim has always been refifted by the bifhops of Down.

PATRONAGE of both fees.—Six parifhes in each diocefe, 12 in the whole, including the *two* deanries, are in the gift of the Crown, 3 are in the Lord Primate, viz. 2 in Down, and 1 in Connor; in the bifhop are 53, viz. 15 in Down and 38 in Connor; and in lay hands, there are 14 in Down, and 22 in Connor.

The church of LISBURN was by the patent of James I. conftituted the cathedral for the united bifhopricks of Down and Connor: but an act of parliament having been paffed for reftor-

ing

ing the cathedral of Down in the city of *Downpatrick*, it is actually repairing in a ſtyle of Gothic architecture, conformable to the venerable remains of the ancient building. That of Connor remains a ruin. There is no epiſcopal palace in either thoſe dioceſes.

---

5. BISHOPRICK of DERRY.

THIS ſee was conſtituted in the year 1158 : its greateſt length is 47, and its greateſt breadth 43 miles *, extending into four counties.

Counties:	Acres.	Pariſhes.	Benefices.	Churches.	Glebe-houſes.	Glebes only.	Benefices without glebe.	Rectories improp.	Wholly improp.
Londonderry	284,100	27	22	25	17	8	-	1	-
Donegal	139,300	10	10	13	7	2	1	-	-
Tyrone	233,100	11	11	13	9	2	-	-	-
Antrim	2,500	part	—	—	-	-	-	-	-
Total	659,000	48	43	51	33	12	1	1	-

The Chapter conſiſts only of a Dean, an Archdeacon, and 3 Prebendaries.

Patronage. That of the Crown includes 3 pariſhes which are the corps of the deanry ; that of the biſhop 33 ; the univerſity of Dublin 3 ; and lay patrons preſent to 9.

There are in the city of *Londonderry* a pretty good cathedral, and a large palace erected by Dr. Barnard, the preceding biſhop. This city is but 32 miles diſtant from the remoteſt part of the dioceſe.

* This dioceſe extends in length 6o Engliſh miles, and in breadth 54!.

6. BISHOPRICK of RAPHOE.

IT is not precifely known at what time this fee was founded, but it muſt have been prior to the tenth century, ſince biſhops of Raphoe are mentioned in the ninth. This dioceſe compriſes the greater part of Donegal, being 44 miles in length from N. to S. and 32 in breadth *.

County of Donegal	Acres.	Pariſhes.	Benefices.	Churches.	Glebe-houſes.	Glebes only.	Benefices without glebe.	Rectories improp.	Wholly improp.
	515,250	31	25	32	17	8	-	3	-

The Dean, the Archdeacon, and 4 Prebendaries compoſe the CHAPTER.

The PATRONAGE of 6 pariſhes, which form the corps of the deanry, is in the Crown; of 15 others in the biſhop; of 7 in the univerſity of Dublin; and of 3 in lay hands.

In the ſmall town of Raphoe there is a very neat though not a large cathedral, which ſerves alſo for a pariſh church, and the biſhop's palace is an old, but convenient edifice. This epiſcopal ſee is at one extremity of the dioceſe, and near 40 miles from the other.

---

* Raphoe is 56 miles long, and 40 broad, Engliſh meaſure.

7. BISHOP-

7. BISHOPRICK of CLOGHER.

THE ſee of CLOGHER was founded by St. Patrick, about the ſame time as Armagh. It ſtretches 60 miles from N. W. to S. E. by a breadth of 20 ;* and compriſes ſome portion of *five* ſeveral counties.

Counties.	Acres.	Pariſhes.	Benefices.	Churches.	Glebe Houſes.	Glebes only.	Benefices without glebe.	Rectories improp.	Wholly improp.
Donegal	25,000	1	1	1	1	–	–	–	–
Fermanagh	254,150	15	15	23	8	7	-	–	–
Tyrone	68,100	4	4	5	2	2	-	–	–
Monaghan	179,600	21	20	20	14	6	-	5	–
Louth	1,850	part of two	—	—	1	-	-	–	–
Total	528,700	41	40	49	26	15	-	5	–

The CHAPTER of Clogher conſiſts of a Dean, Precentor, Chancellor, Archdeacon, and *five* Prebendaries.

PATRONAGE. The Crown preſents to *one* pariſh ; the biſhop to 33 ; the univerſity to 4 ; and lay patrons to *two*.

The cathedral of this dioc_se, which is at the ſame time the pariſh church, is a plain, handſome, modern ſtructure. The palace is large, with a remarkable fine park and demeſne. They are both in the *city*, or rather village of *Clogher*, which is about 33 miles diſtant from the fartheſt part of the dioceſe.

* The length of Clogher is 76 Engliſh miles, and the breadth 25.

R 2 S. BISHOP-

8. BISHOPRICK of KILMORE.

THIS fee was founded in the *thirteenth* century, and in the 15th changed its ancient name of *Brefny*, into that of KILMORE. It lies parallel to, and fouth of Clogher, extending 58 miles in length, and from 10 to 20 in * breadth, through *four* counties.

Counties.	Acres.	Parishes.	Benefices.	Churches.	Glebe Houfes.	Glebes only.	Benefices without glebe.	Rectories improp.	Wholly improp.
Fermanagh	29,300	3	2	3	-	3	—	—	--
Leitrim	184,750	10	7	8	1	6	—	—	--
Cavan	281,000	26	26	25	8	16	—	10	--
Meath	2,200	part	part	—	-	—	—	—	--
Total	497,250	39	30	36	9	25	—	10	--

There are a Dean and an Archdeacon, but no *Chapter* in this diocefe.

PATRONAGE. *Three* parifhes, which are confolidated in the deanry, are in the Crown; 33 in the bifhop; *one* in the univer-fity; and 2 in lay patrons.

There is no cathedral, and the parifh church of *Kilmore* is very fmall and ancient. It joins the bifhop's palace, which is a large houfe fituated in a fine demefne, about 3 miles from Cavan, and 39 from the remoteft part of the diocefe.

* The length of Kilmore is 74 Englifh miles, and the breadth from 13 to 25.

3   9. BISHOP-

9. BISHOPRICK of ARDAGH.

THE fee of ARDAGH was founded in the middle of the 5th century. In 1658 it was united to the bishoprick of *Kilmore*, but in 1692 Dr. Ulysses Burgh was appointed to it separately. Upon his death, in the same year, it was re-united to *Kilmore*, and continued fo, till Dr. Hort was promoted from those fees to the archbishoprick of Tuam in 1741, when they were again separated, and *Ardagh* was annexed to the archbishoprick; which union has continued ever since, though the diocese of *Elphin* intervenes between them. It extends into 6 counties, and yet it is a very small diocese; the greatest length from N. to S. being 42 miles, and the breadth, which is in some places but 4, never exceeding 14 *.

Counties.	Acres.	Parishes.	Benefices.	Churches.	Glebe Houses.	Glebes only.	Benefices without glebe.	Rectories improp.	Wholly improp.
Cavan	10,600	3	part	3	-	1	-	3	-
Leitrim	71,200	7	7	8	3	4	-	2	-
Sligo	4,400	1	part	1	-	-	-	1	-
Roscommon	8,700	1	1	1	-	-	1	1	-
Longford	129,850	22	14	15	6	8	-	9	-
Westmeath	8,900	3	2	1	1	-	1	2	1
Total	233,650	37	24	29	10	13	2	18	1

* The length of *Ardagh* is 61 English miles, and the breadth from 5 to 18.

In

In this diocefe there are a Dean and Archdeacon, but no CHAP-
TER nor epifcopal refidence ; neither can the church of *Ardagh*
be called a cathedral.

The PATRONAGE of *one* parifh belongs to the Crown ; the
bifhop prefents to 30 ; and a lay patron to 6.

---

10. BISHOPRICK OF MEATH. *

SEVERAL fmall bifhopricks gradually coalefced into one fee,
which received the name of MEATH, at the end of the 12th
century : and in 1568 the bifhoprick of Clonmacnois was incorporated with it by act of parliament. It extends from the fea to the
Shannon, over part of fix counties, in length from E. to W. 63
miles, and in breadth about 20 † at a medium.

Counties.	Acres.	Parifhes.	Benefices.	Churches.	Glebe Houfes.	Glebes only.	Benefices without glebe.	Rectories improp.	Wholly improp.
Meath	324,400	147	59	44	19	33	15	38	24
Cavan	9,400	1	1	1	1	—	—	—	—
Longford	4,300	1	1	1	-	1	1	—	—
Weftmeath	222,750	59	31	20	6	14	14	14	7
King's Co.	102,000	16	7	11	3	2	2	12	4
Kildare	750	part of one.	—	—	-	—	—	—	—
Total	663,600	224	99	77	29	51	32	64	35

There is no cathedral in this diocefe : neither is there a CHAP-
TER or even a Dean of Meath ; the only dignities are the deanry

* The bifhop of Meath has precedence of all other bifhops.
† The diocefe of Meath extends in length 80, in breadth 25 Englifh miles.

Clonmacnoife,

of *Clonmacnoife*, and the archdeaconry of *Meath*. The want of a chapter is fupplied by a *Synod*, of which every incumbent is a member, and the archdeacon prefident; their proceedings are authenticated by a common feal.

PATRONAGE. The Crown prefents to 81 parifhes; the lord Primate to 2; the bifhop to 69 and the deanery; and 37 are in lay patronage.

The epifcopal refidence at *Ardbraccan*, near the town of Navan, is a large and convenient manfion, erected by the prefent bifhop, in a ftyle of fuperior elegance, and yet with fuch fimplicity, as does equal honor to his Lordfhip's tafte and liberality. It ftands about 46 miles from the S. W. extremity of the bifhoprick.

THE PROVINCE OF DUBLIN CONTAINS FIVE SEES UNDER FOUR PRELATES.

---

1. DUBLIN.

THE firft mention we find of this fee is in the 7th century. In the year 1152 it was erected into an *archbifhoprick:* and in 1214 the bifhoprick of GLANDALOUGH, which had been founded in the 6th century, was incorporated with DUBLIN.

It is 50 miles in length from N. to S. and 36 in the greateft breadth *; containing the whole county of *Dublin*, the moft of *Wicklow*, and part of two others.

* The length of this diocefe in Englifh meafure is 64 miles, the breadth 46.

5

Counties.	Acres.	Parishes.	Benefices.	Churches.	Glebe-houses.	Glebes only.	Benefices without glebe.	Rectories impr.op.	Wholly impr.op.
Dublin	142,050	100	54	56	23	7	23	13	1
Kildare	75,000	57	12	9	4	8	1	9	2
Wicklow	257,400	49	19	16	7	7	5	1	-
Wexford	2,900	2	1	1	1	—	—	—	-
Queen's Co.	600	1	part of union	—	—	—	—	—	-
Total	477,950	209	86	82	35	22	29	23	3

The CHAPTER of *St. Patrick's*, Dublin, confifts of the Dean, (who is elective by the archbifhop and chapter) the Precentor, Chancellor, Treafurer, 2 Archdeacons, of *Dublin* and of *Glandelough*, and 19 Prebendaries.

The members of the *collegiate* CHAPTER of *Chrift-church* are the Dean, Precentor, Chancellor, Treafurer, the Archdeacon of Dublin, and 3 Prebendaries. This deanry is annexed to the bifhoprick of *Kildare*.

The PATRONAGE of the diocefe is thus divided: The Crown prefents to 15 parifhes; the archbifhop to 144; the lord chancellor and the three chief judges, in conjunction with his grace, to 2; the chapter of Chrift church, or its members feparately, to 23; the chapter of St. Patrick, and its members, to 9; and lay-patrons to 16 parifhes.

The cathedral of St. Patrick is a large and venerable pile; and the archbifhop's palace a very antique building. His grace has alfo a fmall country refidence at *Tallagh* of no very modern date. The moft remote part of the diocefe is 34 miles from Dublin.

2. KILDARE.

2. KILDARE.

THE bishoprick of Kildare was founded about the end of the 5th century. Its greatest length from E. to W. does not exceed 36 miles, nor its greatest breadth 23 *. This see comprehends part of three counties.

Counties.	Acres.	Parishes.	Benefices.	Churches.	Glebe houses.	Glebes only.	Benefices without glebe.	Rectories improp.	Wholly improp.
Kildare	161,000	56	27	14	6	9	12	24	1
King's Co.	121,000	18	9	8	1	1	7	3	—
Queen's Co.	49,000	7	5	6	2	1	2	4	—
Total	332,200	81	31	28	9	11	21	31	1

In the CHAPTER of this diocese there are a Dean, Precentor, Chancellor, Treasurer, Archdeacon, 4 Prebendaries, and 4 Minor Canons.

PATRONAGE. The Crown is patron of 27, the bishop 30, and different laymen of 24 parishes.

The cathedral is small, but kept in good order, and is also the parish church : but the bishop has no place of residence in his diocese, and is always dean of Christ church, Dublin.

* In English miles 46 by 29.

S 3. FERNS,

3. FERNS, and 4. LEIGHLIN.

These bishopricks were united in the year 1600.

FERNS was founded in 598, and extends about 46 miles in length from N. to S. and 18 in breadth, comprising the whole county of *Wexford*, and a small part of *Wicklow*.

LEIGHLIN was established in 632. It comprehends the entire county of *Carlow*, a considerable part of the *Queen's County*, and extends into *Wicklow* and *Kilkenny*. This diocese is of a very irregular form; in some places but 6, and in none above 13 miles broad, though it is 39 * miles long from N. to S.

The union extends 62 by 25+ miles.

FERNS.

Counties.	Acres.	Parishes.	Benefices.	Churches.	Glib. H uses.	Glebes only.	B nefices without Glebes.	Rectories improp.	Wholly improp.
Wexford	340,000	141	38	39	4	27	13	45	13
Wicklow	12,200	2	2	1	–	1	1	1	–
Total.	352,200	143	40	40	4	28	14	46	13

LEIGHLIN.

Counties.	Acres.	Parishes.	Benefices.	Churches.	Glib. H uses.	Glebes only.	B nefices without Glebes.	Rectories improp.	Wholly improp.
Wicklow	42,000	7	2	2	–	2	1'	1	—
Carlow	137,050	49	17	13	1	4	10	11	—
Queen's Co.	122,000	27	16	13	1	4	11	6	—
Kilkenny	17,850	6	4	2	–	2	2	2	—
Total.	318,900	89	39	30	2	12	24	20	—
Total of union	671,100	232	79	71	6	40	38	66	13

* The breadth of Leighlin is from 8 to 16, the length about 50 English miles.

+ In English miles, nearly 79 by 32.

The

The *Chapter* of FERNS confists of a Dean, Precentor, Chancellor, Treafurer, Archdeacon, and 10 Prebendaries.

The *Chapter* of LEIGHLIN is compofed of the fame dignitaries, with only 4 Prebendaries.

The PATRONAGE of the 232 parifhes, in this union, ftands thus. The Crown prefents to 3 in *Ferns*, and 15 in *Leighlin*: The bifhop to 115 in *Ferns*, and 56 in *Leighlin*; the univerfity of Dublin to *one* in *Leighlin*; and lay-patrons to 25 in *Ferns*, and in *Leighlin* to 12: the title to five others in this diocefe is in litigation.

The cathedral of FERNS is fmall, and quite plain: that of LEIGHLIN, though not large, is built in the form of a cathedral at *Old Leighlin*, and very neat: they both ferve for parifh churches.

At FERNS there is a handfome and convenient palace, erected by the late bifhop Cope, and completely finifhed by the prefent bifhop. It is 33 miles diftant from the fartheft part of FERNS, and 42 from the extremity of LEIGHLIN.

---

5. OSSORY.

This fee, which was originally eftablifhed at *Saigair*, and afterwards at *Aghavoe*, was founded very early in the 5th century: it includes almoft the whole of *Kilkenny*, a good part of the *Queen's County*, and fome of the *King's County*, extending 36 miles in length, from N. to S. and 23 * in breadth.

* In Englifh miles the length is 46, and the breadth 29.

The

Counties.	Acres.	Parishes.	Benefices.	Churches.	Glebe-houfes.	Glebes only.	Benefices without glebe.	Rectories improp.	Wholly improp.
Kilkenny.	281,900	120	45	28	11	27	12	31	1
Queen's Co.	60,000	15	10	7	4	6	-	5	-
King's Co.	4,100	1	1	1	-	-	1	1	-
Total	346,000	136	56	36	15	33	13	37	1

The CHAPTER is formed by the Dean, Precentor, Chancellor, Treafurer, and Archdeacon, with 7 Prebendaries.

PATRONAGE. In the Crown are 26 parifhes, 76 in the bifhop, 4 in the dean or chapter, and 30 in lay patrons.

The cathedral is a large handfome pile, dedicated to *St. Canice*, whence the borough of Irifhtown, in which it ftands, (adjoining to the city of *Kilkenny*) derives its name. The bifhop has a good houfe clofe to the cathedral, which is fituated about 30 miles from the fartheft part of the diocefe.

THE PROVINCE OF CASHEL COMPREHENDS ELEVEN SEES, UNDER THE ARCHBISHOP AND FIVE SUFFRAGANS.

1. CASHEL AND 2. EMLY.

THE fee of CASHEL was either founded or reftored at the beginning of the 10th century ; and was made an *archbifhoprick* in 1152. EMLY, which was founded in the 5th century, is faid to have been at firft an Archbifhoprick alfo. They were united in 1568.

The

The archbishoprick is almost confined to the county of *Tippe-rary*, branching a very little way into *Kilkenny* and *Limerick*, and is 28 miles in length, and 23 in breadth. EMLY, which is 32 miles long, from N. to S. and about 12 broad, comprises a part of *Tipperary*, and a larger scope in *Limerick*. The united fees * are very compact, extending 32 miles one way, and 30 the other.

CASHEL.

Counties.	Acres.	Parishes.	Benefices.	Churches.	Glebe Houses.	Glebes only.	Benefices without glebe.	Rectories impop.	Wholly improp.
Tipperary	276,550	96	27	23	10	10	8	11	3
Limerick	850	1	Part	—	—	—	-	—	-
Kilkenny	600	Part	Part	—	—	—	-	—	-
Total	278,000	97	27	23	10	10	8	11	3

E M L Y.

Tipperary	51,900	20	6	4	3	3	2	3	—
Limerick	86,150	38	14	8	3	6	4	14	—
Total	138,050	58	20	12	6	9	6	17	—
Total of union	416,050	155	47	35	16	19	14	28	3

The CHAPTER of *Cashel* consists of a Dean, Precentor, Chancellor, Treasurer, and Archdeacon, with 4 Prebendaries. The CHAPTER of *Emly* contains the like number of dignities and Prebends, the *treasurership* excepted.

The cathedral of CASHEL, which serves also for a parish church, is a large and handsome edifice, compleated by the present

* In English measure the length of Cashel is 35 and the breadth 29 miles; of *Emly* 41 and 15. The whole union is 41 by 38.

Archbishop; the ancient and spacious church which, still venerable in ruins, stands upon the rock of Cashel, having fallen to
decay in the time of his predecessor. The palace is a plain, large
house, in the city of Cashel, to which a public library is annexed.
There is no part of these united Sees more distant than 25 miles
from CASHEL.

3. & 4. WATERFORD and LISMORE.

Thefe fees were united in the year 1536: the bishoprick of
LISMORE had been founded in the beginning of the 7th century; but that of WATERFORD was not established till the 11th,
when the *Oftmen* settled on the Irish coast. This very small diocefe is confined to the eastern part of the county of *Waterford*,
and does not extend above 10 miles in length and 7 * in breadth.
But the diocese of LISMORE is 30 miles long and about 29
broad, including the greatest part of *Waterford county* and a confiderable portion of *Tipperary*. The union stretches in length 39
miles from E. to W. and 29 in breadth.

WATERFORD.

Counties.	Acres.	Parishes.	Benefices.	Churches.	Glebe Houses.	Glebes only.	Benefices without glebes.	Rectories improp.	Wholly improp.
Waterford	31,300	33	9	8	2	4	6	2	3
LISMORE.									
Waterford	231,500	41	20	14	5	10	5	26	-
Tipperary	92,000	32	15	8	1	17	3	13	-
Total	323,500	73	35	22	6	27	8	39	-
Total of union	354,800	106	44	30	8	31	14	41	3

* The dimenfions of *Waterford* are nearly 13 by 9; of *Lifmore* 38 by 37; and of the union,
49 by 37 Englifh miles.

The

The CHAPTER of WATERFORD confifts only of the Dean, Precentor, Chancellor, and Treafurer, for there are no Prebendaries, and the Archdeacon has not a vote. But in LISMORE the Archdeacon is a member of the CHAPTER with the 4 other dignitaries and 10 Prebendaries.

PATRONAGE. In *Waterford* 12 parifhes are in the gift of the Crown, and 16 in that of the bifhop. In *Lifmore* the Crown prefents to 12, the bifhop to 27, and lay patrons to 30. The remainder are wholly impropriate.

The cathedral of *Waterford* is a very elegant church erected within a few years. The bifhop's palace, which ftands clofe by it, is a handfome modern ftructure of no great fize, and very much confined in fituation. At *Lifmore* the cathedral is fpacious and handfome. There are fome parts of *Lifmore* 35 miles diftant from the city of Waterford.

---

5. CLOYNE.

The bifhopric of CLOYNE was founded in the 6th century. It was united to Cork for upwards of two hundred years : but it has now continued feparate above a century. It lies entirely within the county of *Cork*, extending E. and W. near 50 miles in length, by a breadth of 23 *.

County.	Acres.	Parifhes.	Benefices.	Churches.	Glebe-houfes.	Glebes only.	Benefices without glebes.	Rectories improp.	Wholly improp.
Cork	539,700	137	69	51	5	43	25	30	14

* Cloyne is 63 Englifh miles long by 29 broad.

The

The Chapter is very full, being compofed of a Dean, Chancellor, Treafurer, Archdeacon and 14 Prebendaries.

In the Patronage of the crown, there are 10 parifhes, in the bifhop's 106; in lay patrons 7: two are in difpute, and 11 are wholly impropriate. The wardenfhip of the church of *Youghal*, which is collegiate, is perpetually united to the bifhoprick.

The cathedral is a fine old building accompanied by a round tower; and is alfo the parifh church. The bifhop's palace is a plain convenient houfe, with good gardens and demefne; but 40 miles from the weftern bounds of the diocefe.

---

6 & 7.   CORK and ROSS.

The foundation of the bifhoprick of Cork is placed in the 7th century; that of Ross is unknown: they were united by queen Elizabeth in 1586, are both contained in the county of *Cork*, and are partly intermixed. The diocefe of Cork is 58 miles long from E. to W. and about 13 broad. The length of the principal part of Ross is 25 miles from E. to W. and the breadth 6: the detached part of it, in the mountains of Bear and Bantry, is about the fame length, but not more than 4 miles broad. The whole union is about 65 by 17 *.

* In Englifh meafure the length of Cork is 74 miles, and the breadth 16; of Rofs 32 by 8; and of the union 83 by 23.

CORK.

County.	Acres	Parishes.	Benefices.	Churches.	Glebe Houses.	Glebes only.	Benefices without glebe.	Rectories improp.	Wholly improp.
Cork	356,300	94	49	41	9	32	15	11	10

ROSS.

Cork	124,000	33	15	12	5	11	3	13	1
Total of union	480.300	127	64	53	14	43	18	24	11

The CHAPTERS of thefe diocefes confift each of a Dean, a Precentor, a Chancellor, a Treafurer, and an Archdeacon; there are befides 12 Prebendaries in *Cork*, and 5 in *Rofs*.

The PATRONAGE of thefe united fees is thus diftributed. The Crown prefents to 7 in *Cork* and *one* in *Rofs*; the bifhop to 66 in the former, and 28 in the latter; lay patrons to 14 in the two, 11 being wholly impropriate.

The *Cathedral* of St. Finbarry is a plain modern church. Near it ftands the bifhop's palace, a large new houfe, built but a few years, by the late bifhop, Dr. Mann. It is 50 miles from the city of *Cork*, to the remoteft parts of ROSS.

---

8 & 9. LIMERICK, ARDFERT, AND AGHADOE.

THE bifhoprick of LIMERICK was united in 1663 to thofe of ARDFERT and AGHADOE, which had long been fo incorporated,

T

as to form but one diocefe.   ARDFERT or ARDART was eftablifhed in the 5th century, and LIMERICK before the 13th.   The
latter extends in length from E. to W. 27 miles in the county of
Limerick, and 17 in breadth ; taking in a fmall part of *Clare*.
The bifhoprick of ARDFERT, which includes the whole county
of *Kerry*, and a portion of *Cork*, is 52 miles long from N. to S.
and 48 broad.   There are 71 miles from one extremity of the
union to the other *.

LIMERICK.

Counties.	Acres.	Parifhes.	Benefices.	Churches.	Glebe Houfes.	Glebes only.	Benefices without glebes.	Rectories improp.	Wholly improp.
Limerick	294,450	85	45	25	9	28	17	15	—
Clare	12,500	3	2	1	1	—	1	1	—
Total	306,950	88	47	26	10	28	18	16	.

ARDFERT AND AGHADOE.

Kerry	647,650	83	40	20	3	7	33	32	3
Cork	28,800	5	1	1	1	-	—	5	-
Total	676,450	88	41	21	4	7	33	37	3
Total of union	983,400	176	88	47	14	35	51	53	3

The CHAPTER of *Limerick* is complete, having the five greater
dignities and 11 prebends.   In that of *Ardfert*, there are no p r-
bends ; but the Archdeacon of *Aghadoe* has a ftall.   This dignity

---

* The dimenfions of thefe bifhopricks, in Englifh meafure, are as follows :---*Limerick* 34 by
21 miles. *Ardfert* 66 by 61. The whole length 90 miles.

and

and the ruined walls of a church at *Aghadoe*, with a round tower, are all the memorials of the bishoprick that now remain.

PATRONAGE. The Crown presents to 8 parishes in *Limerick*, and to 19 in *Ardfert*; and the bishop to 46 in the former and 38 in the latter; 13 parishes in *Limerick* depend on the chapter, and 52 in both upon lay patrons.

The cathedral of *Ardfert* is no more than an old parish church; but that of *Limerick* is an ancient and venerable pile. The bishop's palace is a comfortable modern house, pleasantly situated on the Shannon at the west end of the city; about 30 miles from the bounds of *Limerick*, and 80 from some parts of *Ardfert*.

---

10 & 11. KILLALOE and KILLFENORA.

THE diocese of KILLALOE was founded early in the 5th century; in the 12th it was incorporated with the ancient bishoprick of *Roscrea* founded in 620; and in the year 1752 the see of KILLFENORA, which had been established about the 12th century, was united to it; and though very small in extent and value, had continued separate till after the restoration, when it was first annexed to the archbishoprick of *Tuam*: that union continued 81 years till 1741, when *Ardagh* being annexed to Tuam, this bishoprick was given in commendam to the bishop of Clonfert.

The diocese of KILLALOE stretches * 80 miles in length, thro' the counties of *Clare* and *Tipperary*, into the *King's County*, and includes also a small part of the *Queen's County*, *Galway*, and *Limerick*. It varies in breadth from 7 to 25 miles.——KILLFENORA is confined to the baronies of *Burrin* and *Corcomroe*, and extends only 18 miles by 9 *.

* The bishoprick of *Killaloe* extends upwards of 100 English miles in length, its breadth is from 9 to 32. The dimensions of *Kilfenora*, are 23 by 11 miles.

T 2 KILLALOE.

KILLALOE.

Counties.	Acres.	Parishes.	Benefices.	Churches.	Glebe-houses.	Glebes only.	Benefices without glebes.	Rectories improp.	Wholly improp.
Clare	426,700	57	20	15	1	15	6	16	5
Tipperary	134,500	41	15	12	1	15	6	5	-
King's Co.	50,000	16	5	6	1	9	1	3	-
Queen's Co.	3,200	1	Part	-	-	-	-	-	-
Galway	8,800	2	1	1	-	-	1	-	-
Limerick	5,300	2	1	1	-	1	-	-	-
Total	628,500	119	42	35	3	40	14	24	5

KILLFENORA.

Clare	37,000	19	8	3	1	5	2	-	2
Total of union	665,500	138	50	38	4	45	16	24	7

In the CHAPTER of each diocefe there are ftalls for a Dean, Precentor, Chancellor, Treafurer, and Archdeacon; and in that of *Killaloe* for 5 Prebendaries.

Of the PATRONAGE of thefe diocefes it is difficult to form an abftract: the rectories being moftly feparate from the vicarages, and many of them in lay patronage. Thus multiplied in number, 10 of them are in the gift of the Crown; 131 in the bifhop; and 36 in lay patrons: thefe 177 rectories and vicarages are united and condenfed, if the expreffion may be allowed, into 50 benefices.

The church of *Killaloe* is not large for a cathedral, but venerable for its antiquity, and in good prefervation, though built above 660 years. It ferves like many others for the parifh church.

Very

Very near the little town of *Killaloe*, in the midſt of a fine de-meſne, beautifully ſituated on the weſtern bank of the Shannon, ſtands the epiſcopal reſidence, a handſome new houſe, erected by the preſent Archbiſhop of Dublin, when biſhop of Killaloe. This ſee is 50 miles from the S. W. extremity of the dioceſe.

THE PROVINCE OF TUAM COMPRISES SIX DIO-CESES, OVER WHICH THE ARCHBISHOP AND THREE SUFFRAGANS PRESIDE.

---

1. TUAM.

This biſhoprick was eſtabliſhed early in the 6th century, and is conſiderably the largeſt in the kingdom, extending over a great part of the counties of *Galway* and *Mayo*, and including a part of *Roſcommon*. It is upwards of 60 miles long, and 50 broad *. With this ſee the biſhoprick of ARDAGH has been held in com-mendam by the preſent archbiſhop and his three predeceſſors.

* The extent of Tuam is 77 by 63 Engliſh miles.

Counties.	Acres.	Parishes.	Benefices.	Churches.	Glebe Houses.	Glebes only.	Benefices without glebes.	Rectories improp.	Wholly improp.
Galway	675,250	49	11	12	1	2	9	1	-
Mayo	424,700	37	10	11	1	2	7	1	-
Roſcommon	35,700	3	3	1	-	-	2	-	-
Total	1,135,650	89	23	24	2	4	18	2	-

The CHAPTER conſiſts of a Dean, a Provoſt, an Archdeacon, and 8 Prebendaries.

PATRONAGE. The deanry alone is in the Crown; 79 pariſhes are in the biſhop's gift; and 10 are united to the Wardenſhip of Galway. The conſtitution of that large and ancient collegiate church is unique in this kingdom; the warden and 3 aſſiſting vicars being elected by the mayor and corporation; the warden annually, and the vicars for life. In the little city of _Tuam_ there is a very neat but ſmall cathedral, which is alſo the pariſh church. The biſhop's palace is a large antique fabric; from which no part of this extenſive diocefe is 50 miles diſtant.

---

2 & 3. CLONFERT AND KILLMACDUAGH.

The former of theſe ſees was founded, near the cloſe of the 6th century; and the latter in the beginning of the 7th. They were united in 1602. CLONFERT lies chiefly in the county of _Galway_; a ſmall part only of _Roſcommon_ belongs to it, with a ſingle

3   pariſh

parifh, on the eaft of the Shannon, in the *King's County*. The greateft length of this diocefe is 29, and the greateft breadth 25 miles. KILLMACDUAGH is wholly in *Galway*, and meafures 18 by 12 miles. The extent of the Union is about 37 by 25 *.

Counties.	Acres.	Parifhes.	Benefices.	Churches.	Glebe Houfes.	Glebes only.	No Glebes.	Rectories improp	Wholly improp.
Galway	193,100	37	10	9	2	12	1	-	-
Rofcommon	17,500	2	1	1	-	2	-	-	-
King's Co.	4,400	1	Part	-	-	—	-	-	-
	215,000	40	11	10	2	14	1	-	-
KILLMACDUAGH.									
Galway	64,000	20	4	4	-	8	-	-	-
Total of union	279,000	60	15	14	2	22	1	-	-

The members of the CHAPTER of *Clonfert*, are a Dean, an Archdeacon, a Sacrift, and 8 Prebendaries. In that of *Killmac-duagh* there are, a Dean, a Provoft, a Chancellor, an Archdeacon, and 2 Prebendaries.

The PATRONAGE of the deanery belongs to the Crown, 31 parifhes to the bifhop, and 9 to a lay patron, in *Clonfert*. In *Killmacduagh* the Crown prefents to 3, the bifhop to 12, and a lay patron to 5.

The cathedral and parifh church of *Clonfert* are the fame: near

* The dimenfions, reduced to Englifh miles are—of *Clonfert*, 37 by 32 ; of *Killmacduagh* 23 by 15; of the Union, 47 by 32.

them

them ſtands the biſhop's palace; there not being one at *Clon-fert*, which is diſtant 21 miles from the fartheſt part of the diocefe, and 34 from the weſtern extremity of the Union. Of the cathedral of *Killmacduagh* nothing but the walls remain, which ſtand near the ruins of a monaſtery, and ſeveral chapels. A large round tower of very ancient and rude maſonry denotes the antiquity and the former confequence of this now wretched hamlet.

---

4. ELPHIN.

This fee dates its origin from St. Patrick in the middle of the 5th century. It compriſes the greater part of the county of *Roſcommon*, a large fcope in *Sligo* and *Galway*, and a very little in *Mayo*. In length, from N. to S. it extends 63 miles, but in breadth it varies from 2 to 24 *.

Counties.	Acres.	Parishes.	Beneficet.	Churches.	Glebe-houſes.	Glebes only.	No Glebes.	Rectories impropr.	Wholly impropr.
Roſcommon	284,650	51	23	20	2	4	17	31	1
Sligo	87,700	16	4	3	1	1	2	8	-
Galway	48,800	8	2	3	2	—	—	7	-
Mayo	1,000	part	—	—	—	—	—	—	-
Total	420,150	75	29	26	5	5	19	46	1

The CHAPTER confiſts of a Dean, Precentor, Archdeacon, and 3 Prebendaries.

* Elphin is about 30 Engliſh miles long, varying in breadth from 3 to 30.

PATRON-

PATRONAGE. The Crown prefents to 2 parifhes, the bifhop to 72, and a lay-patron to *one*.

The cathedral, which is alfo the parifh church is neither large nor fplendid : but the bifhop's palace is a very good modern houfe, in the midft of an excellent demefne, and adjoining the fmall town of *Elphin*, which is about 35 miles from the northern boundary of the diocefe.

5 & 6. KILLALLA and ACHONRY.

The bifhopric of KILLALLA was founded about the fame time as Elphin, and in the following century the fee of ACHONRY was eftablifhed. They both extend into the counties of *Mayo* and *Sligo*; the river *Moy* and the *Ox Mountains* forming the boundary between them. The greateft length of *Killalla* is from E. to W. 45 miles, by a breadth of 21. *Achonry* ftretches from N. E. to S. W. 28 miles, and is 21 broad *. The United Sees meafure E. and W. 55 miles, and from N. to S. 21.

KILLALLA.

Counties.	Acres.	Parifhes.	Benefices.	Churches.	Glebe Houfes.	Glebes only.	No Glebes.	Rectories improp.	Wholly improp.
Mayo	271,200	17	6	6	2	2	2	-	—
Sligo	43,100	8	5	6	4	1	-	2	—
Total.	314,300	25	11	12	6	3	2	2	—

ACHONRY.

Counties.	Acres.	Parifhes.	Benefices.	Churches.	Glebe Houfes.	Glebes only.	No Glebes.	Rectories improp.	Wholly improp.
Mayo	93,700	13	3	2	1	-	2	13	—
Sligo	113,950	14	6	6	1	3	2	9	—
Total.	207,650	27	9	8	2	3	4	22	—
Total of union	521,950	52	20	20	8	6	6	24	—

* The length and breadth of *Killalla*, in Englifh miles, are—57 by 27; of *Achonry*, 35 by 27; and of the Union, 70 by 21.

U

The CHAPTER of *Killalla* is compofed of a Dean, Precentor and Archdeacon, with 5 Prebendaries. In that of *Achonry* there are the fame dignities, with only 3 prebends.

PATRONAGE. The Crown prefents to 2 parifhes, which are the corps of the refpective deanries, in each of thefe diocefes: the other 48 parifhes are in the gift of the bifhop.

The cathedral of KILLALLA is fmall, but venerable for its antiquity: it is the only church in the parifh, though a round tower at the other end of the village indicates the ancient fite of another church, of which however no veftige remains.

The palace is a very fmall and ruinous houfe, ill fituated and ill contrived, at the edge of a very fine demefne; but the prefent bifhop is making fuch additions and improvements as will render it a very comfortable refidence. It is not 30 miles diftant from the moft remote part of either diocefe.

Abftract of the ECCLESIASTICAL ESTABLISHMENT.

IN this abftract the table is augmented by two additional columns: one of them exhibits the proportion of *acres* to the number of *churches*; and the other fhews the proportion which *impropriations* bear to the whole number of parifhes, in every diocefe.

N. B. The diocefes are placed in each province according to their refpective fize; and the figures prefixed to them point out their comparative extent with refpect to the whole number.

Dioceses.	Acres.	Parishes.	Benefices.	Churches.	Average Acres to each Church.	Glebe Houses.	Parishes with Glebe only.	Benefices without Glebes.	Rectories improp.	Parishes wholly impropriate.	Proportion of Impropriations.
5 Meath	663,600	224	99	77	8,618	29	51	32	64	35	2.26
6 Derry	659,000	48	43	51	12,921	33	12	1	1	—	—
7 Down&Connor	597,450	114	65	76	7,861	23	17	26	17	15	3.56
9 Clogher	528,700	41	40	49	10,789	26	15	—	5	—	8.20
11 Raphoe	515,250	31	25	32	16,000	17	8	—	3	—	10.33
12 Kilmore	497,250	39	30	36	13,812	9	25	—	10	—	3.90
16 ARMAGH	468,550	103	69	69	6,761	51	13	14	12	9	4.90
22 Ardagh	233,650	37	24	29	8,056	10	13	2	18	1	1.94
23 Dromore	155,800	26	24	27	5,770	14	2	8	3	—	8.66
Prov. of Armagh	4,319,250	663	419	446	9,684	212	156	83	133	60	3.43
3 Ferns and Leiglin	671,100	232	79	71	9,545	6	40	38	66	13	2.93
15 DUBLIN	477,950	209	86	82	5,828	35	22	29	24	1	8.64
20 Ossory	346,000	136	56	36	9,351	15	33	13	37	1	3.57
21 Kildare	332,200	81	31	28	11,864	8	11	21	31	1	2.53
Prov. of Dublin,	1,827,250	658	252	217	8.3°1	64	106	101	158	16	3.82
2 Limerick and Ardfert	983,400	176	88	47	20,923	14	35	51	52	3	3.20
4 Killaloe and Kilfenora	665,500	138	50	38	17,513	4	45	16	24	7	5.11
8 Cloyne	539,700	137	69	51	10,582	5	43	25	30	14	3.11
14 Cork and Ross	480,300	127	64	53	9,062	14	43	18	24	11	3.62
17 CASHEL and Emly	416,850	155	47	35	11,887	16	19	14	28	3	5.00
19 Waterford & Lismore	354,800	106	44	30	11,826	8	31	14	41	3	2.40
Prov. of Cashel,	3,439,750	839	362	254	13,542	61	216	138	199	41	3.49
1 TUAM	1,135,650	89	23	24	47,318	2	4	18	2	—	44.5
10 Killala and Achonry	521,950	52	20	20	26,097	8	8	6	24	—	2.16
18 Elphin	420,150	75	29	26	16,159	5	5	19	46	1	1.59
21 Clonfert and Kilmacduagh	279,000	60	15	14	19,928	2	22	1	—	—	—
Prov. of Tuam.	2,356,750	276	87	87	29,249	17	39	44	72	1	3.75
Total of the Kingdom	11,943,000	2436	1120	1101	11,910	354	517	366	562	118	3.58

* Lough Neagh contains 58,200 acres, not included in any diocese, which being added to 11,943,000 makes 12,001,200 acres; the number returned in page 17.

It appears by the foregoing pages, that there are in the whole kingdom 2436 *parishes*, which form at present 1123 *benefices*, with cure of Souls; exclusive of 111 sinecures in the several dioceses.

It appears also, that the *churches* amount to 1001; and the glebe-houses to 354. Of the 1123 benefices there are only 366 *destitute of glebe*.

The table shews how large a proportion of the tythes are *impropriate*, or the property of laymen. In the diocese of Elphin, they are in the ratio of *two* to *three*, and taking the whole kingdom as 1 to 3.58 which is *two sevenths* of the whole.

═══════════

IV. ROUND TOWERS.

Whatever difference there be in the opinions of antiquaries concerning the use of the round towers peculiar to Ireland; it is universally agreed, and indeed their situation being always near a church proves, that they were erected for some religious purpose. A more compleat list of them, than what has yet appeared, may be acceptable to the curious.

─────────

In ULSTER, 9.

County of Antrim,   near *Antrim, *at a place called Steeple.*
                         at Armoy.
                         in *Ram Island, *in Lough Neagh.*

4   *County*

County of Cavan,	at Drumlane.
County of Down,	at *Drumbo.
	at *Maghera, *half only is standing.*
County of Fermanagh,	in *Devenish Ifland, *in Lough Erne.*
County of Monaghan,	at Clones.
	at *Enifkeen.

In LEINSTER, 24.

County of Dublin,	at *Clondalkin.
	at *Lufk.
	at Rathmichael.
	at *Swords.
County of Kildare,	at *Caftledermot.
	at *Killcullen.
	at *Kildare.
	at Oughterard.
	at Taghadoe.
County of Kilkenny,	at *Aghavuller, *part only remaining.*
	at Fertag, alias *Beggar's Inn.*
	at *Kilkenny, or *rather in Irifhtown.*
	at Kilree.
	at Tulloherin.
King's County.	at *Clonmacnoife—two.
County of Louth,	at *Dromifkin, *part of one.*
	at *Monafterboyce.
County of Meath,	at *Donaghmore.
	at *Kells.
Queen's County,	at *Dyfert.
	at *Timahoe.
County of Wicklow,	at Glandelough—two, *one of them perfect.*

In MUNSTER, 14.

County of Clare,	at *Drumcliff.
	at *Dyfart.
	in Inifcalthra, *in Lough Deirgeart.*
	in Scattery Ifle, *in the Shannon.*
County of Cork,	at *Cloyne.
	at *Ballybeg, *the ſtump of one.*
	at Kineth.
County of Kerry,	at *Aghadoe.
	at Rattoo.
County of Limerick,	at *Dyfert.
	at *Kilmallock.
County of Tipperary,	at *Cafhel, *on the Rock.*
	at *Rofcrea.
County of Waterford,	at Ardmore.

In CONNAUGHT, 9.

County of Galway,	at Feartamore.
	at *Kilmacduagh.
County of Mayo,	at Aghagower.
	at Ballagh.
	at *Killalla.
	at *Melick.
	at *Turlough.
County of Rofcommon,	at Oran.
County of Sligo,	at *Drumcliffe, *the ruin of one.*

* Thofe marked with an Afterifk have been feen by the Author.

Thefe

Thefe 56 towers are all ftanding ; there were five others ftill perfeᵪ within a few years.

1. Co. Down, at *Downpatrick*, lately taken down.*
2. Co. Dublin, in *Ship-ftreet*, DUBLIN, deftroyed a few years fince.
3. Co. Cork, at *Cork*, pulled down about fifty years.
4. Ditto, at *Brigown*, not long down.
5. Co. Kerry, at *Ardfert*, which fell in 1770.

V. POSTSCRIPT.

THE communication of fome official documents, which were laid before the Irifh parliament, while this book was in the prefs, tempts me to add to it a few pages ; with a more correᵪt account of the population, and fome particulars concerning the prefent ftate of the agriculture, manufaᵪtures, and exports of Ireland.

The increafe of population has been aftonifhingly rapid: the number of inhabitants having been *trebled* in little more than a century.   Soon after the revolution they were eftimated by Sir William Petty, at no more than 800,000.   In the year 1695, a computation was made from the returns of the colleᵪtors of hearth-money ;

---

* The round tower of *Downpatrick* was taken down, in order to enlarge the weft end of the cathedral, which is now repairing, after having lain in ruins for a great number of years. And it is very remarkable, that under the foundations of this tower, were found the vefliges of a more ancient church, which appears to have been of exceeding good mafonry, and upon a larger fcale than the prefent old fabrick, in the walls of which there are many pieces of cut flone, that have evidently been ufed in fome former building. The fame circumftance may alfo be obferved in feveral of the ruined churches at *Clonmacnoife*.

3   by

by which it appears, that they amounted, at that period, to fomething more than *a million*, there being then 200,000 houfes. I
find that the number of houfes returned by thofe officers at Lady-
day 1781 was 477,602 *.   An account was laid before the Houfe
of Commons in the courfe of this feffion †; in which the
houfes of each county are claffed, according to the number of
hearths they contain, and amount in the whole to 701,102 ‡.
And if we allow only *five* perfons to a houfe, the number of inhabitants muft exceed THREE MILLIONS AND A HALF: but
when we take into confideration the great populoufnefs of the pooreft cottages, the many crouded houfes in Dublin and other large
towns, and that the univerfity, the barracks, hofpitals, and public
offices are not included in the hearth-money returns; we may
perhaps, without exaggeration, rate the average number of perfons
at *five and a half* to a houfe, and confequently ftate the population
of Ireland, at this day, to be 3,850,000. This prodigious increafe of population in *one hundred years*, is doubtlefs, in a great
meafure owing to the progreffive improvements in agriculture and
manufactures; fince the moft induftrious counties are the moft
populous.   But we muft not overlook the natural caufes which
have alfo contributed to it : fuch are the mildnefs of the climate,
the abundance and convenience of fuel, and the habits of the
people; who, content with fimple food, are plentifully fupplied
with a wholefome and cheap fuftenance, in that invaluable root the
potatoe.   Certain it is that the culture of the potatoe has increafed
amazingly, in the laft thirty years : and it is as certain, that population invariably follows, where plenty of fubfiftence occurs.

* In the Rev. Mr. Howlett's Effay on the Population of Ireland, printed in 1786.

† This account is dated the 22d of March 1793, and figned *Thomas Wray*, Infpector-Gene-
ral of Hearth-money.

‡ Of this number 112,556 are returned as belonging to paupers, and therefore exemp
from the tax.

                                                      That

That there is employment for this augmented number of inhabitants, appears from the rifing wages of the labourer and artificer, in moft parts of the kingdom. That manufactures and hufbandry are not only much extended in the places where they have been long eftablifhed, but fpreading even into the moft remote counties, has been partly fhewn in the foregoing pages, and is confirmed by the fubjoined extracts of official papers. That this country is in a ftate of high and increafing profperity, is evident from the comparative ftatement of the exports of different periods : I have ftated only two, but were there room here for further extracts concerning the ftate of commerce in former times, the progrefs it has made would be ftill more manifeft. But the late great increafe of the linen manufacture is owing to the export bounties, which commenced in 1781 : as the advancement of tillage is to be dated from 1784. * Bounties on exportation had operated with various fuccefs at different times : but it was in that year that Mr. Fofter, now Speaker of the Houfe of Commons, framed the excellent fyftem of regulations, which have rendered bounties effectual.

Some judgment may be formed of the ftate of tillage, from an account of the mills which are folely appropriated to the grinding of wheat and making of flour †. Of thefe there are 249.

In *Ulfter*, Co. of Monaghan	1			
Leinfter, Co. Kilkenny	37	*Munfter*,	Co. Wicklow -	1
Co. Tipperary	48			
King's County	22		Co. Limerick -	7
Co. Kildare	20		Co. Cork - -	6
Queen's Co.	19		Co. Waterford	4
Co. Weftmeath	12		Co. Clare - -	3
Co. Meath -	11	*Connaught*,	Co. Galway -	26
Co. Wexford	10		Co. Rofcommon	9
Co. Carlow -	9			
Co. Longford	4			249

* The port of *Dublin* is alone excepted from the benefit of thefe bounties on the exportation of corn ; becaufe a bounty is paid on the inland carriage of corn and flour to the metropolis.

† Extracted from the journals of the Houfe of Commons for 1791.

X Extract

EXTRACT of a Report made to the HOUSE of COMMONS, of the Corn, Provifions, and Linen Cloth, exported in the Year 1791, diftinguifhing the feveral Ports.

Ports.	Barrels of Corn.	Cwt. of Flour, Meal, & Bread.	Barrels of Beef.	Barrels of Pork.	Number of live Oxen or Cows.	No. of live Hogs	Cwt. of Butter.	Yards of Linen Cloth.
ULSTER.								
Antrim Belfaft *	18,127	20,528	7,194	7,196	———		15,809	10,684,441
—— Larne	10	760	4	150	146		2,210	31,763
Lond. Londonderry *	--		14		14		4	1,026,156
—— Colerain			3	———	———		438	153,422
Donegal, Killybegs	4,264							
Down, Donaghadee	2,787	6,458	3	·——	22,661		4	23,190
—— Newry *	120	70	2,066	6,347	3,775	3829	10,970	5,001,283
—— Strangford	8,211		9	——	1,470			34,021
LEINSTER.								
Dublin Dublin *	21,305	7,885	26,374	5,410	28		28,624	19,698,285
Louth Dundalk *	10,212	9,473		———	1,363	882	———	17,986
—— Drogheda *	378,007	30,820	188	·——	26		514	1,969,138
Wexford Wexford *	52,227	25,895	112	212	193	684	1,342	209
—— Rofs	8,602	9,940	1,874	1,984			1,334	3,731
Wicklow, Wicklow	3,278				27	4		
MUNSTER.								
Cork Cork *	49,080	22,374	55,525	38,948	·——	13	139,507	1,197,729
—— Baltimore	2,618	400	7	192			96	4,250
—— Kinfale	4,905			798		50		
—— Youghal	29,585	2,443	235	1,360			3,935	14,135
Kerry, Dingle			8			30	2,252	1,309
Limerick Limerick *	150,464	24,190	10,193	11,661			9,401	12,016
Waterford Waterford	214,971	33,301	12,702	19,660	22	310	78,681	14,135
CONNAUGHT.								
Galway Galway *	500	——	20	260				4,000
Mayo, Newport	4,745		20					
Sligo, Sligo	22,728	2,620	644	724			1,449	61,041
Totals	567,747	170,869	117,196	94,506	29,625	5802	295,575	39,647,246

* Thefe are the only ports into which tobacco can be imported; neither can wine, tea, coffee, or fpirits be admitted into any other, without a particular licence from the commiffioners of the revenue, except *New Towns* and *New Rofs*.

I   AVERAGE

AVERAGE VALUE OF EXPORTS.

	In the 7 years ending Mar. 25, 1777.	In the 7 years ending Mar. 25, 1791.	Increase.	Decrease.
Corn, meal, flour, bread	£.64,871	£.415,645	£.350,774	——
Barrelled beef	312,967	236,000	——	76,967
Barrelled pork	128,435	134,684	6,249	——
Live stock	20,668	151,000	130,332	——
Butter	607,907	591,782	——	16,125
Linen cloth	1,390,919	2,183,514	792,595	——
Yarn *(linen)*	188,810	182,668	——	6,142
	£.2,714,577	£.3,895,293	£.1,279,950	£.99,234

Increase on balance   £.1,180,716

Thus we see, that the commerce of Ireland has been raised upwards of A MILLION annually, upon the abovementioned articles, in the short space of 14 years; and that neither the provision trade nor the exportation of yarn have diminished, in proportion to the advance in the quantities of corn and linen; nay, that if we take the export of live stock into the account, it has increased; notwithstanding the great additional consumption, which so considerable an accession of wealth and population must occasion.

The annual value of all the exports of Ireland, amounted on an average of the last *seven* years to 4,357,000 *l*.

VI. GLOS-

VI. GLOSSARY,

Or, Explanation of some of those Irish words which most frequently occur, in composition with the names of places.

AGH, a Field.
ANAGH or ANA, a River.
ARD, a high Place, or rising Ground.
ATH, a Ford.
AWIN, a River.
BALLY or BALLIN, a Town, or inclosed place of habitation.
BAN or BANE, White, or Fair.
BEG, Little.
BEN, the summit of a Mountain, generally an abrupt head.
BUN, A Bottom, Foundation, or Root.
CAR or CAHIR, a City.
CARRICK, CARRIG, CARROW, a Rock or Stony Place.
CORK, CORCAGH, a Marsh, or swampy Ground.
CLARA, a Plain.
CROAGH, CROGHAN, a sharp pointed Hill resembling a Rick.
CLOGH, CLOUGH, a great Stone.
CURRAGH, a marshy or fenny Plain.
CLON, a Glade, or a level Pasture Ground.
COL, CUL, a Corner
DERRY, a clear dry Spot in the midst of a woody swamp.
DON, a Height or Fastness, a Fortress.
DONAGH, a Church.
DROM, a high narrow ridge of Hills.
INCH, INIS, an Island.
KEN, a Head.
KILL, a Church or Cemetery.

KNOCK,

KNOCK, a fingle Hill, or a Hillock.

LICK, a flat Stone.

LOUGH, a Lake, or a Pool.

MAGH, a Plain.

MAIN, a Collection of Hillocks.

MORE, large, great.

RATH, a Mount or Entrenchment, a Barrow.

ROSS, a point of Land projecting into Waters.

SHAN, Old.

SLIEBH, a range of Mountain, a Hill *covered with Heath*.

TACH, a Houfe.

TEMPLE, a Church.

TOM, TOOM, a Bufh.

TRA, a Strand.

TOBAR, TUBBER, a Well or Spring.

TULLAGH, a gentle Hill, a Common.

TULLY, a Place fubject to floods.

PRINTED BY T. RICKABY.

1792.

INDEX TO THE MAP.

The Names of those Cities, Towns, and Boroughs, which send Representatives to Parliament are printed in SMALL CAPITALS.

The DIOCESE is added to the Names of *Parishes* only ; and the Letter R. denotes the Parish to be a *Rectory*, V. a *Vicarage*, C. a *Curacy*, Ch. a *Chapelry*.

N. B. When the Name of a Town or Village occurs in the *Second* Column, it shews that the Church of the Parish to which it is annexed, is at such Town or Village.—*Ex. gr.* the Church of *Abbeystrowry* is at *Skibbereen*.

Names.	Description.	County.	Barony.		Diocese.
Abbeyodorney	*Village*	Kerry	*Clanmaurice*		
Abbeyfeale	*Village*	Limerick	*Conello*	V.	Limerick
Abbeygormagan		Galway	*Longford*	V.	Clonfert
Abbeylaragh		Longford	*Granard*	V.	Ardagh
Abbeyleix		Queen's Co.	*Cullinagh*	V.	Leighlin
Abbeymahon		Cork	*Barrioe*	R.	Rofs
Abbeyfhrule		Longford	*Shrewle*	R.	Ardagh
Abbeyftrowry	Skibbereen	Cork	*Carbery*	V.	Rofs
Abington	*Village*	Limerick	*Owneybeg*	R.	Emly
Acharrow	*Village*	Sligo	*Carbury*		
Achill	*Island*	Mayo	*Burrifhoole*		
Achil-beg	*do.*	do.	*do.*		
Achil-head	*Promontory*	do.	*do.*		
ACHONRY	*Bishoprick*	Sligo & Mayo			
Achonry		Sligo	*Leeny*	R.	Achonry
Acton	*Village*	Armagh	*Orior*		
Adamftown		Wexford	*Bantry*	R.	Ferns
Adare	*Village*	Limerick	*Cofhma*	V.	Limerick
Adar	*River*	Mayo	*Gallen*		
Addergool		do.	*Tirawly*	V.	Killalla
Addergool		Galway	*Dunnamore*	V.	Tuam
Adnith		Tipperary	*Eliogurty*	V.	Cafhel
Adragool	*Cataract*	Cork	*Bear & Bantry*		
Affane		Waterford	*Defies weathout*	V.	Lifmore
Aghaborg		Monaghan	*Dartree*	R.	Clogher
Aghabulloge		Cork	*Mufkerry*	R.	Cloyne
Aghacrew		Tipperary	*Kilnemana*	R.	Cafhel
Aghacrofs		Cork	*Condons, &c.*	R.	Cloyne
Aghada		do.	*Imokilly*	R.	Cloyne
Aghade		Carlow	*Ravilly*	C.	Leighlin
Aghaderrick	Loughbrickland	Down	*Upper Iveagh*	V.	Dromore
AGHADOE	*Bishoprick*	Kerry			
Aghadoe		do.	*do.*	R.	Aghadoe
Aghadown		Cork	*Carbery*	V.	Rofs
Aghadowy		Londonderry	*Coleraine*	R.	Derry

Names.	Description.	County.	Barony.	Diocese.
Ballymacky		Tipperary	Upper Ormond	R. Killaloe
Ballymacormack		Longford	Ardagh	R. Ardagh
Ballymacward		Galway	Tiaquin	V. Clonfert
Ballymacwilliam		King's Co.	Warrenstown	R. Kildare
Ballymadun		Dublin	Balruddery	V. Dublin
Ballymaganny	Village	Meath	Half Fowre	
Ballymagarvy		Meath	Duleck	V. Meath
Ballymaglasson		Meath	Ratoath	R. Meath
Ballymagorry	Village	Tyrone	Strabane	
Ballymagowran	Village	Cavan	Tullaghah	
Ballymahon	Village	Longford	Rathline	
Ballymakenny		Louth	{ County of Drogheda }	R. Armagh
Ballymany		Kildare	Great Connel	R. Kildare
Ballymartle		Cork	Kinalea, &c.	R. Cork
Ballymascanlan		Louth	Dundalk	C. Armagh
Ballymenah	Town	Antrim	Toome	C. Connor
Ballymodan	Bandonbridge	Cork	Kinalmeaky	V. Cork
Ballymurrin		Tipperary	Eliogurty	V. Cashel
Ballyneale	Village	Kilkenny	Ida, &c.	
Ballynetty	Village	Limerick	Clanwilliam	
Ballynitty		Wexford	Shelmaliere	V. Ferns
Ballypatrick	Village	Tipperary	Iffa and Offa	
Ballyphilip	Portaferry	Down	Ardes	R. Down
Ballyquillane		Queen's Co.	Stradbally	R. Leighlin
Ballyquintin Point	Cape	Down	Ardes	
Ballyragget	Village	Kilkenny	Fassadining	
Ballyrathane		Antrim	Dunluce	V. Connor
Ballysadere	Coloony	Sligo	Leney, &c.	V. Achonry
Ballysadere	Village	Sligo	Tiraghril	
Ballyscadden		Limerick	Small county	R. Emly
Ballyscullen	Ballaghy	Londonderry	Loughinsholin	R. Derry
BALLYSHANNON	Town	Donegal	Tyrhugh	
Ballysheehan		Tipperary	Middlethird	V. Cashel
Ballysonnon		Kildare	Ophaly	R. Kildare
Ballyspellan		Cork	Barrymore	V. Cloyne
Ballyspellin	Village	Kilkenny	Gallmoy	
Ballyvaghan	Bay	Clare	Burrin	
Ballyvaldon		Wexford	Ballaghten	V. Ferns
Ballyvourney		Cork	Muskerry	R. Cloyne
Ballywalter	Village	Down	Ardes	
Ballywillan		Antrim	Dunluce	V. Connor
Ballywire		Armagh	Ferns	C. Armagh
Balrain		Kildare	Ikeath, &c.	R. Kildare
Balroddan		Meath	Deece	R. Meath
BALRUDDERY	Barony	Dublin		
Balruddery	Village	Ditto	Balruddery	V. Dublin
Balsoon		Meath	Deece	R. Meath
Balteagh		Londonderry	Keneght	R. Derry
BALTIMORE	Village	Cork	Carbery	
BALTINGLASS	Town	Wicklow	Talbotstown	R. Leighlin
Ban	Lough	W. Meath	Half Fowre	
Banada	Village	Sligo	Leney	
BANAGHER	Town	King's Co.	Garrycastle	
Banagher		Londonderry	Keneght	R. Derry
Bandon	River	Ditto	Lib. Kinsale	

c

1

d

6

f

D.

5

g

E.

Names.	Description.	County.	Barony.	Diocess.
Eagle-Island	Island	Mayo Coast	Erris	
Earlstown	-	Kilkenny	Shellilogher	R. Offory
Eask—Lough	Lake	Donegal	Boylagh, &c.	
Easterfnew	-	Rofcommon	Boyle	V. Elphin
Edenderry	Town	King's Co.	Cooleftown	
Edermine	-	Wexford	Ballagheen	R. Ferns
Edgeworthftown	Village	Longford	Ardagh	
Effin	-	Limerick	Coshma	R. Limerick
EGLISH	Barony	King's Co.		
Eglifh	-	Ditto	Eglifh	V. Meath
Eglifh	Village	Tyrone	Dungannon	
Eglifh	-	Armagh	Tyranny	C. Armagh
Eightmile Bridge	Village	Down	Upper Iveagh	
Eirke	-	Kilkenny	Gallmoy	R. Offory
ELIOGURTY	Barony	Tipperary		
ELPHIN	Bifhoprick	Rofcommon, Sligo, &c.		
Elphin	Town	Ditto	Rofcommon	R. Elphin
Ematris	Kilcrow	Monaghan	Dartree	R. Clogher
Emlaghrafh	Peninfula	Mayo Coast	Erris	
EMLY	Bifhoprick	Tipperary and	Limerick	
Emly	Village	Tipperary	Clanwilliam	V. Emly
Emlyfadd	Ballymote	Sligo	Corran	V. Achonry
Emlygrennan	-	Limerick	Coshlea	R. Limerick
Emyvale	Village	Monaghan	Trough	
Ennel—Lough	Lake	W. Meath	Moyafhel, &c.	
ENNIS	Town	Clare	Iflands	
Ennifcoffey	-	W. Meath	Fartullagh	R. Meath
ENNISCORTHY	Town	Wexford	Scarewalfh	V. Ferns
Enifkeen	-	Cavan	Clonchee	C. Meath
Ennifkerry	Village	Wicklow	Rathdown	
ENNISKILLEN	Town	Fermanagh	Tyrckenedy	R. Clogher
Ennifnag	-	Kilkenny	Gallmoy	R. Offory
Ennifrufh	-	Londonderry	Loughinfholin	Ch. Derry
Enorelly	-	Wicklow	Arklow	V. Dublin
Erne—Lough	Lake	Fermanagh		
Erne	River	Cavan	Tullohonoho	
Errigall	Garvagh	Londonderry	Coleraine	R. Derry
Errigall	-	Monaghan	Trough	V. Clogher
Errigalkeeroge	-	Tyrone	Clogher	R. Armagh
ERRIS	Barony	Mayo		
Erry	-	Tipperary	Middlethird	R. Cafhel
Efker	-	Dublin	Half Rathdown	V. Dublin
Efky	Village	Sligo	Tyreragh	V. Killalla
Ettagh	-	King's Co.	Ballibritt	R. Killaloe
Eyrecourt	Town	Galway	Longford	

F.

Faghalftown	-	W. Meath	Half Fowre	V. Meath
Faghy	-	Galway	Longford	V. Clonfert
Fahan	-	Donegal	Inifhowen	R. Derry

h

Names.	Description.	County.	Barony.	Diocese.
Fairhead - - -	Cape - - - -	Antrim - -	Cary	
Faithleg - - -	- - - - -	Waterford -	Gualtiere - -	R. Waterford
Fallen - - -	River - - -	Longford -	Longford	
Fanlobbish - - -	Dunmanaway -	Cork - - -	Carbery - -	V. Cork
Farahy - - -	- - - - -	Cork - - -	Fermoy - -	R. Cloyne.
FARBILL - - -	Barony - -	W. Meath		
Farneybridge - -	Village - - -	Tipperary -	Killnalongurty	
Fartagh - - -	Johnstown - -	Kilkenny -	Gallmoy - -	R. Offory
Fartrey - - - -	River - - -	Wicklow -	Newcastle	
FARTULLAGH -	Barony - -	W. Meath		
FASSACHDINING	Barony - -	Kilkenny		
Faughan - - -	River - - -	Londonderry -	Tyrekerin	
Faughanvale - -	Muff - - -	Londonderry -	Ditto - -	R. Derry
Feacle - - -	Village - - -	Clare - - -	Tullagh - -	R.V. Killaloe
Feal - - - -	River - - -	Kerry - - -	Iraghticonner	
Feanagh - - -	- - - - -	Leitrim -	Leitrim - -	R. Ardagh
FEATHERD - -	Village - - -	Wexford -	Shelburne - -	R. Ferns
Fedamore - - -	- - - -	Limerick -	Small County -	V. Limerick
Feighcullen - -	- - - - -	Kildare -	Ophaly - -	R. Kildare
Fenard - - -	Village - - -	Donegal -	Tyrhugh	
Feno - - -	Lake - - -	Leitrim -	Carrigallen	
Fenix - - -	River - - -	Cork - -	Imokilly	
Fennagh - - -	- - - - -	Carlow - -	Idrone - - -	R.V. Leighlin
Fennor - - -	- - - - -	Tipperary -	Slewardagh -	R. Cashel
Fennor - - -	- - - - -	Meath - -	Duleek - -	R. Meath
Fenoagh - - -	- - - - -	Waterford -	Upperthird -	R. Lismore
Fenoagh - - -	- - - - -	Tipperary -	Lower Ormond	R.V. Killaloe
Ferbane - - -	Village - - -	King's Co. -	Garrycastle	
Fergus - - -	River - - -	Clare - - -	Islands	
FERMANAGH -	County - -	Ulster		
FERMOY - -	Barony - -	Cork		
Fermoy - - -	Village - - -	Ditto - - -	Condons, &c.	
Fern—Lough -	Lake - - -	Donegal - -	Kilmacrenan	
FERNS - - -	Bishoprick - -	Wexford, &c.		
Ferns - - -	Village - -	Wexford - -	Scarewalsh -	V. Ferns
FERRARD - -	Barony - -	Louth		
Ferriters - - -	Islands - -	Kerry - - -	Clanmaurice	
FETHARD - -	Town - - -	Tipperary -	Middlethird -	R. Cashel
FEWS - - -	Barony - -	Armagh		
Fews - - -	Village - -	Ditto - - -	Fews	
Fews - - - -	- - - - -	Waterford -	Decies without	V. Lismore
Fiddown - - -	- - - -	Kilkenny -	Iverk - -	R. Offory
Fieldtown - - -	Village - - -	Dublin - -	Nethercrosse	
Fina - - -	River - - -	Monaghan		
Finae - - -	Village - - -	W. Meath -	Half Fowre	
Finglas - - -	Village - - -	Dublin - -	Nethercross -	V. Dublin
Finglas - - -	- - - -	King's Co. -	Clonlisk - -	R. Killaloe
Finn—Lough -	Lake & River -	Donegal - -	Boylagh, &c.	
Finogh - - -	- - - - -	Clare - - -	Bunratty - -	R.V. Killaloe
Fintona - - -	Village - - -	Tyrone -	Clogher	
Fintown - - -	Village - - -	Donegal -	Boylagh, &c.	
Fintra - - -	Bay - - -	Ditto - -	Ditto	
Finuge - - -	- - - - -	Kerry - - -	Clanmaurice -	V. Ardfert
Finvarra Point -	Cape - - -	Clare - - -	Burrin	
Finvoy - - -	- - - - -	Antrim -	Kilconway -	R. Connor
FIRCAL - - -	Barony - - -	King's Co.		

G.

Names.	Description.	County.	Barony.	Diocess.
Gara—Lough	Lake	Sligo	Coolavin	
Gardenhill		Fermanagh	Clonawly	Ch. Clogher
Gare		Tipperary	Slewardagh	V. Cashel
Garfinagh		Kerry	Corcaguinny	V. Ardfert
GARRICASTLE	Barony	King's Co.		
Garricloyne		Cork	Muskerry	R. Cloyne
Garrilough	Village	Wexford	Ballagheen	
Garrison	Village	Fermanagh	Magheraboy	
Garristown	Village	Dublin	Balruddery	V. Dublin
Garrivoe		Cork	Imokilly	V. Cork
Gartan	Churchhill	Donegal	Kilmacrenan	R. Raphoe
Garvagh	Village	Londonderry	Coleraine	
Garvaghy		Down	Lower Iveagh	V. Dromore
Gawnagh—Lough	Lake	Longford	Granard	
Geal		Tipperary	Middlethird	R. Cashel
Geevach	Mountains	Leitrim and Roscommon		
Geneva (New)	Village	Waterford	Gualtiere	
Gernanstown		Meath	Slane	R. Meath
Gernanstown		Louth	Ardee	R. Armagh
Geron Point	Cape	Antrim	Glenarm	
Gerranekennif		Cork	Imiskilly	R. Cloyne
GESHIL	Barony	King's Co.		
Geshil	Village	Ditto	Geshil	R. Kildare
Gessigo Point	Cape	Sligo	Carbery	
Giants Causeway	Promontory	Antrim	Cary	
Gilberstown		Carlow	Forth	R. Leighlin
Gilford	Village	Down	Lower Iveagh	
Gilly—Lough	Lake	Sligo	Carbery	
Giltown		Kildare	Naas	C. Dublin
Girly		Meath	Kells	V. Meath
Glanbane		Tipperary	Clanwilliam	R. Emly
Glanbehy		Kerry	Iveragh	R. Ardfert
Glandelough		Wicklow	Ballinacor	V. Dublin
Glandore	Harbour	Cork	Carbery	
Glanduff	Village	Limerick	Connello	
GLANEHIRY	Barony	Waterford		
Glanely	Village	Wicklow	Newcastle	Ch. Dublin
GLANEROUGHT	Barony	Kerry		
Glanevy	Village	Antrim	Massarecn	V. Connor
Glaninagh		Clare	Burrin	R. Kilfenora
Glankeen	Burroseleagh	Tipperary	Ileagh	V. Cashel
Glanmire	Village	Cork	Barrymore	
Glanmire—Upper	Village	Ditto	Ditto	
Glanmore	Village	Kilkenny	Ida, &c.	
Glanore	Glanworth	Cork	Fermoy	R. Cloyne
Glanton	Village	Ditto	Duhallow	
Glanworth	Village	Ditto	Fermoy	
Glaslough	Town	Monaghan	Trough	
Glass—Lough	Lake	W. Meath	Half Fowre	
Glasscarrick Point	Cape	Wexford	Ballagheen	
Glassnevin	Village	Dublin	Coolock	C. Dublin
Glenaa	Mountains	Kerry	Dunkerron	
GLENARM	Barony	Antrim		
Glenarm	Village	Ditto	Glenarm	
Glencolmkill		Donegal	Boylagh, &c.	R. Raphoe
Glenegad Head	Cape	Ditto	Inishowen	
Glenely	River	Antrim	Glenarm	

i

H.

Names.	Description.	County.	Barony.		Diocese.
Hackettstown	Village	Carlow	Ravilly	R.	Leighlin
Hag's Head	Cape	Clare	Corcomroe		
Haggardstown	- - -	Louth	Dundalk	C.	Armagh
Hainstown	- - -	Kildare	Salt	C.	Kildare
Hamiltonsbawn	Village	Armagh	Fews		
Hangman's Point	Cape	Cork	Lib. Kinsale		
HARRISTOWN	House	Kildare	Naas		
Harristown	- - -	King's Co.	Insul. Kildare	R.	Kildare
Hawlbowling	Island	Cork Harbour	Kinalea, &c.		
Hazelhatch	Village	Dublin	Newcastle		
Headford	Village	Galway	Clare		
Helwick Head	Cape	Waterford	Decies without		
Hen and Chickens	Mountains	Down	Upper Iveagh		
Heynstown	- - -	Louth	Dundalk	R.	Armagh
HILSBOROUGH	Town	Down	Lower Iveagh	R.	Dromore
Hoaretown	- - -	Wexford	Shelmaleire	R.	Ferns
Hoath—Hill of	Promontory	Dublin	Coolock		
Hoath	Village	Ditto	Ditto	C.	Dublin
Hog Head	Cape	Kerry	Dunkerron		
Hollymount	Village	Mayo	Kilmain		
Hollywood	Village	Down	Castlereagh	C.	Down
Hollywood	Village	Wicklow	Talbotstown	R.	Dublin
Hollywood	- - -	Dublin	Balruddery	V.	Ditto
Holmpatrick	Skerries	Ditto	Ditto	C.	Ditto
Holycross	Village	Tipperary	Eliogurty	C.	Cashel
Hooke	- - -	Wexford	Shelburne	V.	Ferns
Hooketower	Lighthouse	Ditto	Ditto		
Hore Abbey	- - -	Tipperary	Middlethird	R.	Cashel
Hornhead	Cape	Donegal	Kilmacrenan		
Horse Island	Island	Kerry	Iveragh		
Horseleap	Village	W. Meath	Moycashel		
Horsepassbridge	Village	Wicklow	Talbotstown		
Horseshoe	Rock	Ditto	Arklow		
Hospital	Village	Limerick	Small County	V.	Emly
Hoyle—Lough	Lake	W. Meath	Corkerry		
Hulin Rocks	Rocks	Antrim	Glenarm		
Hyne—Lough	Bay	Cork	Carbery		

I.

IARCONNAUGHT	- - -	Galway	Moycullin		
JAMESTOWN	Village	Leitrim	Dromahaire		
IBAWNE	Barony	Cork			
IBRICKIN	Barony	Clare			
IDA, IGRIN, and IBERCON	Barony	Kilkenny			
IDRONE	Barony	Carlow			
Jerpoint	Kilkenny	Gowran	V.	Ossory
Jerpoint Monast.	- - -	Ditto	Knocktopher	V.	Ditto
IFFA AND OFFA	Barony	Tipperary			

1

k

4

5

n

O.

Names.	Description.	County.	Barony.	Diocese.
Ramor Lough	Lake	Cavan	Castleraghan	
Ramsgrange	Village	Wexford	Shelburne	
Ramhead	Cape	Waterford	Decies within	
RANDALSTOWN	Town	Antrim	Toome	
Rapharn Lough	Lake	Mayo	Burrishoole	
RAPHOE	Bishoprick	Donegal, &c.		
RAPHOE	Barony	Ditto		
Raphoe	Town	Ditto	Raphoe	R. Raphoe
Rathadoe	Village	Donegal	Raphoe	
Ratharkin		Antrim	Kilconway	R. Connor
Rathee		Ditto	Antrim	R. Ditto
Ratine		Meath	Navan	V. Meath
Ratafs		Kerry	Clanmaurice	R. Ardfert
Rath		Carlow	Ravilly	R. Leighlin
Rath		Clare	Inchiquin	R.V.Killaloe
Rathangan	Village	Kildare	Ophaly	R. Kildare
Rathaspick	Dunane	Queen's Co.	Slewmargy	R. Leighlin
Rathaspick		Wexford	Forth	R. Ferns
Rathaspick		W. Meath	Moygoish	C. Ardagh
Rathbirry		Cork	Ibawne	V. Rofs
Rathbeggan		Meath	Ratoath	V. Meath
Rathbourney		Clare	Burrin	R. Kilfenora
Rathboyne		Meath	Kells	Ch. Meath
Rathbride	Village	Kildare	Ophaly	
Rathcharen		Cork	Carbery	R. Cork
Rathconny	Glanmire	Cork	Lib. Cork	R. Cork
Rathconnel		W. Meath	Moyefael, &c.	V. Meath
RATHCONRATH	Barony	W. Meath		
Rathconrath	Village	Ditto	Rathconrath	R. Meath
Rathcool		Kilkenny	Gowran	V. Offory
Rathcool		Tipperary	Middlethird	R. Cashel
Rathcoole	Village	Dublin	Newcastle	V. Dublin
Rathcore		Meath	Moyfurath	V. Meath
Rathcormuck		Waterford	Upperthird	V. Lifmore
RATHCORMUCK	Village	Cork	Barrymore	R. Coyne
RATHDOWN Half	Barony	Dublin		
RATHDOWN Half	Barony	Wicklow		
Rathdowny	Village	Queen's Co.	Upper Offory	V. Offory
Ratndowtan		Cork	Carbery	R. Cork
Rathdrum	Village	Wicklow	Ballinacor	V. Dublin
Rathdrunmin		Louth	Ferrard	R. Armagh
Rathernon		Kildare	Great Connel	R. Kildare
Rathfarne	Village	W. Meath	Farbill	
Rathfarnham	Village	Dublin	Newcastle	C. Dublin
Rathleigh		Meath	Skryne	R. Meath
Rathfriland	Down	Down	Upper Iveagh	
Rathgoggin	Charleville	Cork	Orrery, &c.	V. Cloyne
Rathjordan		Limerick	Clanwilliam	V. Emly
Rathkeale	Village	Ditto	Connello	R. Limerick
Rathkelty		Tipperary	Eliogurty	R.V.Cashel
Rathkenny		Ditto	Kilnamanna	C. Ditto
Rathkenny		Meath	Navan	V. Meath
Rathkyran		Kilkenny	Iverk	C. Offory
Rathlin	Island	Antrim	Cary	R. Connor
Rathlinan		Tipperary	Clanwilliam	R.V.Cashel

S

r

T.

Names.	Description.	County.	Barony.		Diocese.
Taghfinny	- Village	Longford	Shrowle	R.	Ditto
TALBOTSTOWN	Barony	Wicklow			
Talbotstown	- Village	Ditto	Talbotstown		
Tallagh	- Ditto	Dublin	Newcastle	V.	Dublin
Tallanstown		Louth	Ardee	V.	Armagh
TALLOW	- Town	Waterford	Coshbride	V.	Lismore
Tallowbridge	- Village	Ditto	Ditto		
Tallwater	- River	Armagh	Oneilland		
Tamlaght	- Cough	Londonderry	Loughinsholin	R.	Armagh
Tamlaghtara		Ditto	Kenoght	R.	Derry
Tamlaghtfinlagan		Ditto	Ditto	R.	Ditto
Tamlaghtocrely		Ditto	Loughinsholin	R.	Ditto
Tanderagee	- Town	Armagh	Orior	R.	Armagh
Tankardstown		Limerick	Coshma	R.	Limerick
Tankardstown		Queen's Co.	Ballyadams	R.	Dublin
Taragh	- Village and Hill	Meath	Skryne	V.	Meath
Tara-hill	- Hill	Down	Ardes		
Tara-hill	- Mountain	Wexford	Gorey		
Tarbert	- Village	Kerry	Iraghticonnor		
Tarbert	- Island	Galway coast	Ballinahinch		
Tarmonbarry		Roscommon	Roscommon	R.	Elphin
Tarmonhill	Mountain	Mayo	Erris		
Tartaraghan		Armagh	Oneilland	R.	Armagh
Tascoffin		Kilkenny	Gowran	R.	Offory
Tassaggard	- Village	Dublin	Newcastle	C.	Dublin
Taughboyne		Donegal	Raphoe	R.	Raphoe
Taunagh		Sligo	Tiraghrill	V.	Elphin
Tawney		Dublin	Half Rathdown	C.	Dublin
Tay	- River	Waterford	Decies without		
Teinagh	- Village	Galway	Leitrim	R.	Clonfert
Telltown		Meath	Kells	R.	Meath
Templebodane		Cork	Barrymore	R.	Cloyne
Templeboy		Sligo	Tyreragh	V.	Killalla
Templebredin		Limerick	Coonagh	V.	Emly
Templebreedy		Cork	Kinalea, &c.	V.	Cork
Templebrian		Ditto	Carbery	R.	Ross
Templecairne	- Pettigoe	Donegal	Tyrhugh	R.	Clogher
Templecorran		Antrim	Belfast	V.	Connor
Templecroan	- Cloghanlea	Donegal	Boylagh, &c.	R.	Raphoe
Templederry		Tipperary	Upper Ormond	R. V.	Killaloe
Temple-erry		Ditto	Ikerrin	R.	Cashel
Templeharry		King's Co.	Clonlish	R.	Killaloe
Templehay		Tipperary	Iffa and Offa	V.	Lismore
Templejehally		Tipperary	Arra	R.V.	Emly
Templemaly		Clare	Bunratty	R.V.	Killaloe
Templemartin		Cork	Kinnalmeaky	R.	Cork
Templemichael	- Longford	Longford	Longford	R.	Ardagh
Templemichael		Cork	Kinalea, &c.	R.	Cork
Templemichael		Tipperary	Slewardagh,&c.	R.	Lismore
Templemichael		Wicklow	Arklow	C.	Dublin
Templemore		Donegal	Inishowen	R.	Derry
Templemore	- Village	Tipperary	Eliogurty	V.	Cashel
Templemurry		Mayo	Tirawly	V.	Killalla
Templenecarrigy		Cork	Barrymore	R.	Cloyne
Templeneilan	- Roscommon	Roscommon	Roscommon	V.	Elphin
Templenoe		Tipperary	Clanwilliam	R.	Emly
Templenoe		Kerry	Dunkerron	R.	Ardfert

s

V.

.

www.ingramcontent.com/pod-product-compliance
Lightning Source LLC
Chambersburg PA
CBHW021523210326

41599CB00012B/1365